AF544904

GILKEY'S WEEDS OF THE PACIFIC NORTHWEST

by

La Rea J. Dennis
Assistant Curator of the Herbarium
Oregon State University

Library of Congress Catalog Card Number: 80-82742
ISBN: 0-88246-039-0

Permission has been given by
Oregon State University Press
to use material from their 1957
Weeds of the Pacific Northwest
by Helen M. Gilkey.

INTRODUCTION

In olden times the term weed (weod) apparently was applied to any rank growth of plants, such as a tall, dense stand of grass or shrubby understory, that was difficult to penetrate. As the diversity of crops and areas brought under cultivation increased, there was an increase in the diversity of unwanted and/or harmful plants in cultivated areas. These became known as weeds.

An exact definition of a weed that would apply in all cases is difficult to attain as there are unlimited possibilities. A weed has been considered to be any plant that: (a) is unsightly or troublesome and also useless; (b) is noxious or injurious to crops; (c) grows in excessive, injurious or useless abundance; or (d) is growing where it is not wanted. Everyone has an opinion as to which plants should be considered as weeds, and what is one mans weed may be another mans valuable crop plant. Perhaps we could paraphrase Gertrude Stein and say that "A weed is a weed is a weed".

Occasionally native plants become weeds but most of our weeds have been introduced from other countries or from other areas of the United States. There are several factors which may contribute to the rapid spread of a weed in a new area. It may have been introduced without its insect pests and fungus diseases that help keep it in check in its native area. Also, plants that would normally keep the plant in check through competition in its original range are not present in the new area. Finally, most successful (and hence most detrimental to man) weeds have some special means leading to very high rates of reproduction and spread. These mechanisms may consist of efficient vegetative means of reproduction as rootstocks or stolons or the production of an overabundance of seeds, especially if these seeds can be dispersed by air or spread by animals. Cultivation or other disturbance of the soil enhances the efficiency of these adaptations.

It is impractical in a manual of this type to treat all of the many plants which are or might possibly become problems in our area. The species included are selected from two general groups of weeds. The first consists of weeds widely recognized to be of economic importance in our area. It is hoped that all weeds of this category have been included. The second group of weeds to be included consists of recently introduced species which, for various reasons, show strong possibilities of becoming dangerous economic pests. Some of these species may have only a very local distribution at present, and perhaps other species are omitted that should have been included.

The very nature of weeds makes the establishment of boundaries artificial at best in a book of this type. The area covered is essentially Oregon, Washington and western Idaho. The most important single factor contributing to the variations in climate, and hence the vegetation, in different parts of the area covered is the Cascade Mountain Range. The majority of weeds have a wide distribution, but some are better adapted to the moister, more equable climate west of the Cascades, while others are limited to the drier eastern area. No attempt was made to cite detailed distributional data because, weeds once established in an area tend to increase their range wherever suitable habitats exist, and any exact locality information would soon be out of date.

ACKNOWLEDGEMENTS

This book is a revision and expansion of Weeds of the Pacific Northwest by Helen M. Gilkey which was published in 1957 under joint auspices of the Oregon State University Press and the Agricultural Experiment Station. Permission from the Oregon State University Press to use material from Weeds of the Pacific Northwest is gratefully acknowledged.

A large number of individuals have contributed to this publication. Arnold P. Appleby, Ralph E. Whitesides, Bill D. Brewster, Patrick K. Boren, Larry C. Burrill and Garvin Crabtree all made suggestions of plants to be included. Philis Benson and Julie Kierstead made drawings to add to the ones by Helen M. Gilkey, Cathrine Young, Patricia Packard, Alleda Burlage, Daisy Overlander and Fern C. Duncan. A special thanks goes to Phyllis K. Holmquist for her care in typing the manuscript. I am deeply indebted to Frank H. Smith, Richard R. Halse and Weldon K. Johnston, all of whom read the entire manuscript and made numerous valuable suggestions.

KEY TO FAMILIES AND GENERA

Page

1a Plants not producing flowers or seeds, reproducing by spores.

2a Spores borne on the undersurfaces or margins of the leaves; stems not jointed . . . . (POLYPODIACEAE) *Pteridium* 19

2b Spores borne in terminal cone-like structures; stems jointed . . . . (EQUISETACEAE) *Equisetum* 21

1b Plants producing flowers (these sometimes inconspicuous) and reproducing by seeds.

3a Main veins of the leaves parallel, ultimate branches of the veins not netted.

4a Submerged aquatics; stems weak; leaves usually whorled . . . . (HYDROCHARITACEAE) *Elodea* 25

4b Terrestrial, or if growing in water, not submerged.

5a Sepals and petals present.

6a Flowers brownish or greenish and inconspicuous.

7a Petals 4; leaves in a basal rosette . . . . (PLANTAGINACEAE) *Plantago* 277

7b Perianth parts 6, in 2 sets of 3 . . . . (JUNCACEAE) *Juncus* 59

6b Flowers white or variously colored and showy.

8a Ovary inferior . . . . (IRIDACEAE) *Iris* 66

8b Ovary superior . . . . (LILIACEAE) 61

Robust plants from a thickened rootstock; leaves broad and coarsely veined . . *Veratrum* 61

Plants with a bulb; leaves narrow and grass-like or cylindrical and hollow.

Plants with an odor of onion or garlic; flowers borne in umbels . . . . . *Allium* 61

Plants without an odor of onion or garlic; flowers borne in elongated clusters . . . . *Zigadenus* 65

5b Sepals and petals lacking; flowers subtended by scales or bristles.

9a Leaves 3-ranked; stems often triangular in cross-section . . . . (CYPERACEAE) 57

Stems with basal sheaths, otherwise leafless and bractless; style base usually enlarged and triangular . . . . *Eleocharis* 58

Stems usually leafy or at least with bracts.

Flowers unisexual . . . . *Carex* 58

Flowers with both stamens and pistil present.

Scales of the spikelet 2-ranked . . . . *Cyperus* 57

Scales of the spikelet spirally arranged . . . . *Scirpus* 58

9b Leaves 2-ranked.

10a Inflorescence a dense cylindrical spike; staminate flowers borne above the pistillate flowers . . . . (TYPHACEAE) *Typha* 23

10b Inflorescence various, but not as above; flowers borne in the axils of dry chaffy bracts . . . . (GRAMINEAE) 27

Ligule absent . . . . *Echinochloa* 45

Ligule present.

Ligule mainly a fringe of hairs, sometimes with a membranous base, but the hairs making up more than one-half the length of the ligule.

Spikelets enclosed within a spiny bur . . . . . . *Cenchrus* 55

Spikelets not enclosed within a spiny bur.

Perennials with extensive creeping rootstocks.

Inflorescence a branched panicle; plants 2 to 5 feet tall . . . . *Sorghum* 47

Inflorescence of 4 to 9 spikes; plants more or less creeping, less than 2 feet in height . . . . *Cynodon* 29

Annuals (ours) without extensive rootstocks.

Each spikelet subtended by 1 to several bristles; inflorescence dense and spike-like . . . . *Setaria* 30

Spikelets not subtended by bristles; inflorescence open . . . . *Panicum* 53

Ligule membranous throughout, although sometimes ciliate.

Sheaths partially to completely fused (the margins not merely overlapping).

Leaf-tip boat-shaped . . . . *Poa* 41

Leaf-tip not boat-shaped.

Annuals (ours) . . . . *Bromus* 37

Perennials . . . . *Dactylis* 47

Sheaths open (i.e. not fused) for at least ¾'s of their length, the margins usually overlapping.

Auricles present at least on some leaves.

Spikelets 1 per node.

Spikelets placed edgewise to the main axis . . . . *Lolium* 39

Spikelets placed flatwise to the main axis . . . . *Agropyron* 51

Spikelets 2 or more per node.

Spikelets usually 2 per node, but if 3, then all alike . . *Taeniatherum* 33

Spikelets 3 per node, the middle one sessile and fertile, the lateral ones stalked and sterile . . . . *Hordeum* 35

Auricles absent.

Annuals (ours).

Florets without awns; plants spreading to prostrate, often rooting at the lower nodes.

Spikelets 2-flowered, borne in 2 rows . . . . *Digitaria* 27

Spikelets 3-flowered, crowded . . . . *Sclerochloa* 40

Florets awned; plants erect.

Awn arising from the tip of the lemma.

Spikelets 4- to 12-flowered (ours) . . . . *Festuca* 29

Spikelets (at least the fertile ones) 2-flowered . . . . *Cynosurus echinatus* 53

Awn arising from the back of the lemma.

Robust plants over 1½ feet tall; spikelets large . . *Avena* 49

Delicate plants less than 1 foot tall . . . . *Aira* 55

Perennials.

Inflorescence of paired terminal spikes; spikelets borne in rows on one side of the flattened rachis . . . . *Paspalum* 38

Inflorescence not of paired spikes.

Spikelets with 1 terminal fertile floret and 2 much reduced lower florets.

Inflorescence appearing about as wide as long . . . . *Hordeum jubatum* 35

Inflorescence much longer than wide . . . . *Phalaris* 45

Spikelets not as above.

Inflorescence more or less compact and spike-like.

Spikelets 1-flowered . . . . *Alopecurus* 30

Spikelets 2-4-flowered . . . . . *Cynosurus cristatus* 53

Inflorescence open, never spike-like.

Spikelets borne in pairs, 1 spikelet sessile and fertile, the other stalked and staminate . . . *Sorghum* 47

Spikelets not borne in pairs.

Plants velvety-hairy . . . . *Holcus* 31

Plants not velvety-hairy.

Spikelets 1-flowered . . . . *Agrostis* 45

Spikelets 2-flowered . . . . *Arrhenatherum* 49

3b Leaves netted veined.

11a Petals either free from each other or absent.

12a Stamens more than 10.

13a Sepals 2 . . . . (PAPAVERACEAE) *Eschscholtzia* 121

13b Sepals or calyx lobes 3 or more.

14a Pistils more than 1.

15a Sepals free from each other . . . . (RANUNCULACEAE) 109

Flowers regular . . . . *Ranunculus* 109

Flowers irregular . . . . *Delphinium* 116

15b Sepals fused . . . . (ROSACEAE) 153

Herbs . . . . *Potentilla* 153

Shrubs or vines.

Leaves pinnately compound . . . *Rosa* 154

Leaves palmately compound . . *Rubus* 155

14b Pistil 1.

16a Leaves opposite . . . . (HYPERICACEAE) *Hypericum* 195

16b Leaves alternate . . . . (MALVACEAE) 191

Flowers white or pink . . . . *Malva* 193

Flowers yellow.

Leaves heart-shaped, toothed but not lobed . . . . *Abutilon* 191

Leaves deeply cut into 3 or 5 lobes . . . . *Hibiscus* 191

12b Stamens 10 or less.

17a Herbage with stinging hairs . . . (URTICACEAE) *Urtica* 67

17b Herbage without stinging hairs.

18a Leaves with papery stipules forming a cylindrical sheath around the stem . . . . (POLYGONACEAE) 69

Inner perianth segments increasing in size with age, the outer segments not enlarging; 1 or more of the segments usually bearing a swollen grain-like structure . . . . *Rumex* 69

Perianth segments similar in size and shape, swollen grain-like structures absent . . . . *Polygonum* 73

18b Leaves without papery cylindrical sheathing stipules.

19a Ovary inferior.

20a Aquatics, most of the plant submerged . . . . (HALORAGACEAE) *Myriophyllum* 201

20b Terrestrial, or at least plants not submerged.

21a Sepals 2 . . . . (PORTULACACEAE) *Portulaca* 93

21b Sepals 4 or 5 or obsolete.

22a Flowers in racemes . (ONAGRACEAE) *Epilobium* 199

22b Flowers in umbels . . . . (UMBELLIFERAE) 203

Fruit with a beak several times longer than the body . . . . *Scandix* 204

Fruit with a beak shorter than the body or beakless.

Fruit bristly.

Fruit with a short distinct beak, ribs obscure . . . . *Anthriscus* 203

Fruit beakless, ribs evident.

Bracts at the base of the umbel cleft . . . . *Daucus* 213

Bracts, if any, linear and not cleft . . . . *Torilis* 203

Fruit not bristly.

Flowers yellow . . . . *Pastinaca* 211

Flowers white or greenish.

Fruit distinctly flattened . . . . *Heracleum* 205

Fruit more or less globose.

Styles ½ as long as the fruit; calyx teeth well developed . . . . *Oenanthe* 210

Styles less than ½ as long as the fruit; calyx teeth inconspicuous.

Stems purple-spotted; leaves finely dissected . . *Conium* 207

Stems not purple-spotted; leaves with distinct leaflets . . . . *Cicuta* 209

19b Ovary superior.

23a Flowers irregular.

24a Leaves simple; plants herbaceous . . . . . (VIOLACEAE) *Viola* 197

24b Leaves compound, or if simple then the plants woody . . . . (LEGUMINOSAE) 159

Shrubs.

Plants spiny . . . . *Ulex* 160

Plants not spiny . . . . *Cytisus* 159

Herbs, though sometimes woody at the base.

Leaves of three leaflets.

Leaflets gland-dotted . . . . *Psoralea* 161

Leaflets not gland-dotted.

Flowers in more or less open, one-sided racemes . . . . *Melilotus* 169

Flowers in heads, short spikes or dense racemes.

Fruit kidney-shaped . . . . . *Medicago* 171

Fruit not curved nor kidney-shaped . . . . *Trifolium* 171

Leaflet number other than three.

Leaves palmately compound . . . . *Lupinus* 169

Leaves not palmately compound.

Plants not tendril bearing . . . . . *Astragalus* 162

Plants tendril bearing.

Style not flattened, tipped with a ring of hairs . . . . *Vicia* 167

Style flattened, hairs borne only on 1 side . . . . *Lathyrus* 165

23b Flowers regular.

25a Sepals 4; petals 4; stamens usually 6 (4 long and 2 short) . . . . (CRUCIFERAE) 123

Fruit not splitting open longitudinally to release the seeds.

Fruit broadly winged . . . . *Isatis* 137

Fruit not winged.

Plants bearing stalked glands . . . . *Chorispora* 131

Plants not glandular . . . . *Raphanus* 151

Fruit splitting open longitudinally to release the seeds.

Fruit little or not at all longer than wide.

Fruit with 2 broad wings . . . . *Thlaspi* 133

Fruit not broadly winged.

Seeds 2 or 4 per fruit.

Fruit more or less inflated; perennial from extensive rootstocks; upper leaves with ear-like lobes at the base of the blades . . . . *Cardaria* 129

Fruit somewhat compressed or flattened; plants, if perennial, then without ear-like lobes at the base of the blades of the upper leaves . . . . *Lepidium* 123

Seeds more than 4 per fruit.

Flowers pale yellow . . . . *Camelina* 127

Flowers white.

Leaves all basal (ours) . . . . *Draba* 150

Leaves on the stem in addition to a basal rosette . . . . *Capsella* 132

Fruit at least twice as long as wide.

Flowers white.

Leaves compound . . . . *Cardamine* 139

Leaves simple.

Plants without hairs . . . . *Conringia* 135

Plants with hairs . . . . *Arabidopsis* 150

Flowers yellow.

Leaves entire or merely toothed.

Plants densely hairy . . . . *Erysimum* 137

Plants without hairs . . . . *Conringia* 135

Leaves, at least the lower, deeply lobed, cleft or compound.

Fruit distinctly beaked . . . . *Brassica* 143

Fruit not beaked.

Plants bearing branched (star-shaped) hairs . . . . *Descurainia* 135

Plants without branched hairs.

Upper leaves sessile and clasping; style evident . . . . *Barbarea* 149

Upper leaves stalked (the stalk sometimes short); style obsolete . . . . *Sisymbrium* 141

25b Without this combination of characters.

26a Leaves compound.

27a Shrubs or vines . . . . (ANACARDIACEAE) *Rhus* 188

27b Herbs.

28a Leaves palmately compound with three leaflets . . . . (OXALIDACEAE) *Oxalis* 177

28b Leaflet number other than three.

29a Ovary deeply 5-lobed, 5-beaked . . . . (GERANIACEAE) 173

Fertile stamens 5 . . . . *Erodium* 174

Fertile stamens 10 . . . . *Geranium* 173

29b Ovary not deeply 5-lobed.

30a Leaves opposite . . . . (ZYGOPHYLLACEAE) *Tribulus* 187

30b Leaves alternate . (ROSACEAE) *Sanguisorba* 158

26b Leaves simple.

31a Flowers usually showy, petals present.

32a Ovary 5-lobed . . . . (GERANIACEAE) *Geranium* 173

32b Ovary not 5-lobed.

33a Sepals 2 . . . . (PORTULACACEAE) 91

Flowers borne in the axils of the upper leaves; petals usually rose-red . . . . *Calandrinia* 92

Inflorescence not leafy; flowers white or pink . . . . *Montia* 91

33b Sepals or calyx lobes 4 or 5 . . . . (CARYOPHYLLACEAE) 95

Sepals free from each other.

Petals entire.

Styles usually 5; petals white; leaves appearing whorled . . . . . . *Spergula* 100

Styles 3; petals usually red or pink (ours) . . . . *Spergularia* 107

Petals lobed, notched or fringed.

Styles 5, capsule opening by 10 teeth . . . . *Cerastium* 103

Styles usually 3 or 4, but if 5, then capsule not opening by 10 teeth.

Petals deeply 2-lobed . . *Stellaria* 105

Petals fringed . . . . *Holosteum* 102

Sepals fused, the tube nearly as long or longer than the lobes.

Styles 5 (ours).

Petals with appendages at the junction of the blade and the claw . . . . *Lychnis* 97

Petals unappendaged . . . . *Agrostemma* 107

Styles 2 or 3 (ours).

Styles 3 . . . . *Silene* 95

Styles 2.

Petals with appendages at the junction of the blade and the claw . . . . *Saponaria* 95

Petals unappendaged . . . . *Vaccaria* 99

31b Flowers inconspicuous, petals absent.

34a Plants with milky juice . . . . (EUPHORBIACEAE) 179

Plants densely covered with minute star-shaped hairs . *Eremocarpus* 185

Plants without star-shaped hairs . . . . *Euphorbia* 179

34b Plants without milky juice.

35a Leaves whorled; plants prostrate . . . . (AIZOACEAE) *Mollugo* 90

35b Leaves alternate or opposite.

36a Leaves opposite.

37a Plants fleshy or mealy or both . . . . (CHENOPODIACEAE) *Atriplex* 83

37b Plants neither fleshy nor mealy . . . . (CARYOPHYLLACEAE) *Scleranthus* 101

36b Leaves alternate.

38a Leaves deeply lobed from the apex into oblong segments; delicate herbs . . (ROSACEAE) *Alchemilla* 157

38b Leaves not deeply lobed from the apex.

39a Calyx papery; plants neither woody, fleshy nor mealy . . . .(AMARANTHACEAE) *Amaranthus* 87

39b Calyx greenish; plants either woody, fleshy or mealy . . . . (CHENOPODIACEAE) 79

Leaves spine-tipped.

Leaves fleshy, nearly round in cross-section . . . . *Halogeton* 79

Leaves flattened . . . . *Salsola* 85

Leaves not spine-tipped.

Calyx lobes with hooked spines . . . . *Bassia* 81

Calyx lobes without hooked spines.

Flowers unisexual . . . . *Atriplex* 83

Flowers, or at least most of them, with both stamens and pistil present.

Herbage glandular and/or mealy . . . . . . *Chenopodium* 81

Herbage hairy, but not glandular nor mealy . . . *Kochia* 85

11b Petals united at least at the base.

40a Ovary inferior.

41a Stamens free from each other.

42a Leaves alternate . . . . (CAMPANULACEAE) *Campanula* 286

42b Leaves opposite or whorled.

43a Leaves whorled . . . . (RUBIACEAE) 279

Corolla lobes much longer than the tube . *Galium* 281

Corolla lobes much shorter than the tube . . . . *Sherardia* 279

43b Leaves opposite.

44a Stamens 1 to 3 . . . . (VALERIANACEAE) *Valerianella* 284

44b Stamens 4 . . . . (DIPSACACEAE) *Dipsacus* 283

41b Stamens united to each other.

45a Staminate flowers in racemes, pistillate flowers axillary . . . . (CUCURBITACEAE) *Marah* 285

45b Flowers in heads surrounded by a cluster of bracts (the involucre) . . . . (COMPOSITAE) 289

aa Flowers all strap-shaped, plants with milky juice.

Leaves all basal.

Receptacle with chaffy bracts . . *Leontodon* 292

Receptacle naked.

Outer series of involucral bracts turned downward; heads solitary . *Taraxacum* 291

Outer series of bracts erect; heads usually 2 or more . . . . . *Hypochaeris* 289

Leaves not all basal.

Pappus absent . . . . *Lapsana* 299

Pappus present.

Pappus of short scales; flowers blue . . . . *Cichorium* 305

Pappus of hairs or plumose bristles.

Achenes flattened.

Achenes beaked . . . . *Lactuca* 301

Achenes not beaked . . . . *Sonchus* 293

Achenes nearly cylindrical or more or less angled.

Pappus of plumose bristles; heads large . . . . *Tragopogon* 309

Pappus of thread-like hairs.

Achenes with a circle of scales at the base of the long slender beak . . . . *Chondrilla* 307

Achenes without a terminal circle of scales . . . . *Crepis* 297

aa' At least some of the flowers tubular; plants without milky juice.

bb All of the flowers tubular.

Pappus consisting of scales or absent.

Pistillate involucre bur-like, armed with spines.

Spines hooked . . . . *Xanthium* 309

Spines straight . . . . *Ambrosia* 311

Involucre not bur-like, although the individual involucral bracts often spine-tipped.

Receptacle greatly elongated and cone-shaped . . . . *Matricaria* 351

Receptacle flat or convex.

At least the lower leaves opposite . . . . *Iva* 315

All the leaves alternate.

Shrubs (ours) . . . . *Artemisia* 355

Herbs.

Receptacle densely bristly . . . . *Centaurea* (in part) 331

Receptacle naked . . . . . . *Tanacetum* 353

Pappus consisting of hairs or bristles.

Involucral bracts nearly equal (in one series) or sometimes with a second series of much reduced bracts at the very base of the involucre.

All flowers with both stamens and pistils . . . . *Senecio* (in part) 358

Outer flowers pistillate . . . . *Erechtites* 362

Involucral bracts in several series.

Leaves not spiny or prickly on the margins.

Shrubs . . . . *Chrysothamnus* 327

Herbs.

Involucral bracts hooked at the tip . . . . *Arctium* 327

Involucral bracts not hooked at the tip.

Involucral bracts entire, pearly white . . . . . . *Anaphalis* 357

Involucral bracts either spiny or fringed, never pearly white . . . . *Centaurea* (in part) 331

Leaves spiny or prickly on the margins.

Heads almost concealed by the upper leaves . . . . *Cnicus* 348

Heads not concealed by the upper leaves.

Pappus of plumose bristles . . . . *Cirsium* 339

Pappus merely roughened, not plumose.

Receptacle fleshy and conspicuously honey-combed . . . . *Onopordum* 347

Receptacle neither fleshy nor honey-combed.

Leaves conspicuously white-mottled along the veins . . . . *Silybum* 344

Leaves never white-mottled . . . . *Carduus* 347

bb' Strap-shaped ray flowers present, although sometimes these not showy.

Pappus absent.

Ray flowers yellow.

Involucral bracts wholly or largely enclosing the achene . . . . *Madia* 321

Involucral bracts only partly clasping the achene . *Hemizonia* 323

Ray flowers white, pink, purple or red.

Ray flowers 3-5; heads small, borne in ± flat-topped clusters . . . . *Achillea* 349

Ray flowers usually 10 or more; heads not borne in flat-topped clusters.

Receptacle chaffy; leaves finely dissected . . . *Anthemis* 351

Receptacle naked; leaves entire or merely toothed or lobed.

Leaves essentially basal . . . . *Bellis* 329

Leaves alternate on the stem as well as basal . . . . *Chrysanthemum* 355

Pappus present.

Ray flowers white or purplish, inconspicuous . . . . *Conyza* 325

Ray flowers yellow.

Recptacle naked.

Involucral bracts in a single series or sometimes with a second series of much reduced bracts at the very base of the involucre . . . . *Senecio* (in part) 358

Involucral bracts in several series . . . . *Solidago* 325

Receptacle chaffy.

Leaves all opposite . . . . *Bidens* 319

Leaves, or all but the lowest ones, alternate.

Leaves linear, spine-tipped . . . . *Hemizonia* 323

Leaves broadly ovate, never spine-tipped . . . *Helianthus* 317

40b Ovary superior.

46a Flowers irregular.

47a Stamens 6, in 2 sets of 3; sepals 2; petals 4 . . . . (FUMARIACEAE) *Fumaria* 122

47b Stamens 5 or less.

48a Plants without green leaves; root parasites . . . . (OROBANCHACEAE) *Orobanche* 275

48b Plants with green leaves.

49a Ovary, at maturity, separating into 4 nutlets.

50a Ovary not deeply lobed; corolla only slightly irregular . . . . . (VERBENACEAE) *Verbena* 249

50b Ovary deeply 4-lobed; corolla strongly irregular . . . . (LABIATAE) 239

Anthers 2 . . . . *Salvia* 245

Anthers 4.

Plants conspicuously white-woolly . . . . *Marrubium* 245

Plants not white-woolly.

Corolla scarcely 2-lipped; plants very aromatic . . . . *Mentha* 239

Corolla strongly 2-lipped; plants not aromatic.

Calyx teeth equal or nearly so . . . . . . *Lamium* 240

Calyx teeth clearly unequal.

Plants prostrate; flowers axillary . . . . *Glechoma* 243

Plant erect or nearly so; flowers in a terminal spike . . . . . . *Prunella* 247

49b Ovary not splitting into 4 nutlets, but the fruit splitting open at maturity to release the seeds.

51a Fruit over 2 inches long, horned; plant sticky; leaves rounded . . . . (MARTYNIACEAE) *Proboscidea* 259

51b Fruit less than 1 inch long . . . . (SCROPHULARIACEAE) 261

Anthers 5 . . . . *Verbascum* 261

Anthers 2 or 4.

Anthers 2 . . . . *Veronica* 262

Anthers 4.

Corolla with a slender elongated spur.

Leaves sessile . . . . *Linaria* 270

Leaves stalked . . . . *Kickxia* 267

Corolla without a spur.

Flowers yellow . . . . *Parentucellia* 268

Flowers white, pink or purple.

Flowers about 5/8 of an inch long; leaves 2 inches long or less . . . . *Antirrhinum* 269

Flowers 1½ to 2 inches long; leaves 4 inches to 1½ feet long . . . . . *Digitalis* 266

46b Flowers regular.

52a Stamens 10 or more.

53a Leaves palmately compound with three leaflets; stamens 10 . . . . (OXALIDACEAE) *Oxalis* 177

53b Leaves simple; stamens more than 10 . . . . . . (MALVACEAE) 191

Flowers white or pink . . . . *Malva* 193

Flowers yellow.

Leaves more or less heart-shaped, toothed, but not lobed . . . . *Abutilon* 191

Leaves deeply cut into 3-5 lobes . . . . *Hibiscus* 191

52b Stamens less than 10.

54a Pistils 2, styles united above.

55a Stamens and stigma united, the column bearing hood-like appendages . . . . (ASCLEPIDACEAE) *Asclepias* 221

55b Stamens and stigma not united . . . . (APOCYNACEAE) 217

Stems erect; flowers pinkish or nearly white . . . . *Apocynum* 219

Stems trailing; flowers blue or purple . . . . *Vinca* 217

54b Pistil 1.

56a Stamens borne directly in front of the corolla lobes . . . . (PRIMULACEAE) 215

Flowers yellow or yellow streaked with purplish red . . . *Lysimachia* 251

Flowers orange to salmon colored . . . . *Anagallis* 216

56b Stamens alternating with the corolla lobes.

57a Twining or trailing herbs . . . . (CONVOLVULACEAE) 223

Leafless non-green parasites . . . . *Cuscuta* 223

Leaves green . . . . *Convolvulus* 224

57b Not twining or trailing.

58a Ovary deeply 4-lobed, separating at maturity into nutlets . . . . (BORAGINACEAE) 231

Flowers yellow to orange . . . . *Amsinckia* 231

Flowers white, blue, purple or red, sometimes with a yellow center.

Nutlets with hooked or barbed prickles or bristles.

Nutlets evenly prickly over the back . . . . *Cynoglossum* 235

Nutlets unevenly prickly, the marginal prickles much longer than the dorsal ones . . . . *Hackelia* 237

Nutlets without prickles or bristles.

Fruiting calyx enlarged and conspicuously veined; flowers blue . . . . *Asperugo* 237

Fruiting calyx not enlarged; flowers white to bluish-white.

Plants fleshy and without hairs . . . . *Heliotropium* 235

Plants not fleshy, bearing hairs.

Flowers borne in the axils of the upper leaves . . . *Lithospermum* 235

Inflorescence not leafy . . . . *Plagiobothrys* 233

58b Ovary not deeply 4-lobed and not separating into nutlets.

59a Leaves all basal; flowers brownish, borne in spikes . . . . (PLANTAGINACEAE) *Plantago* 277

59b Leaves not all basal.

60a Stigma capitate or only slightly 2-lobed . . . . (SOLANACEAE) 250

Fruit a dry capsule splitting open at maturity to release the seeds.

Leaves sessile . . . . *Hyoscyamus* 250

Leaves stalked . . . . *Datura* 257

Fruit a fleshy berry.

Berry enclosed in a bladdery-inflated calyx; corolla lobes not reflexed . . . . *Physalis* 257

Berry not completely enclosed by the calyx; calyx not inflated; corolla lobes reflexed . . . . *Solanum* 251

60b Stigmas 2 or 3.

61a Flowers over ½ inch wide (ours) . . . . (HYDROPHYLLACEAE) *Nemophila* 227

61b Flowers less than ½ inch wide (ours) . . . . (POLEMONIACEAE) 229

Leaves and bracts spine-tipped . . . . *Navarretia* 230

Leaves not spine-tipped.

Calyx uniform throughout, not papery between the lobes . . . . *Polemonium* 229

Calyx more or less papery between the lobes . . . . *Gilia* 229

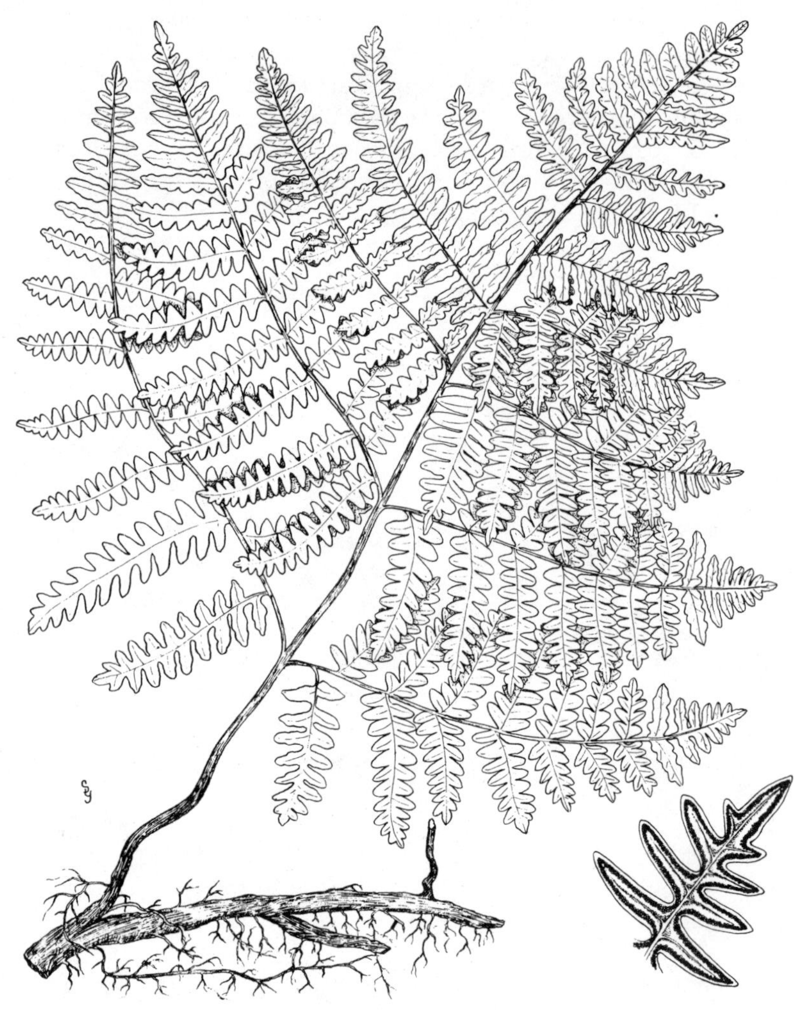

WESTERN BRACKEN
Pteridium aquilinum x 1/12

Fern Family: POLYPODIACEAE

Plants with underground rootstocks; leaves (fronds) developing by unrolling from the tips; reproduction by powdery spores borne in clusters of minute containers on the lower surface or margins of the fronds.

Many species of fern are native to the Pacific Northwest, but only the Western Bracken can be considered a weed of major importance.

WESTERN BRACKEN OR BRACKEN FERN

Pteridium aquilinum (L.) Kuhn var. *pubescens* Undw.

Large coarse ferns, perennial underground by thick dark creeping branching rootstocks; fronds (leaves) large, 1 1/2 to 5 or 6 feet in height, broadly triangular in outline, the lower segments several times compound; stalk shining yellow, its base brown velvety; spores brown, borne abundantly beneath the inrolled margins of leaflets; entire frond turning brown and dying in autumn.

This species is cosmopolitan and is a common fern in the Pacific Northwest, thriving in shade, sun, deep rich soil, heavy clay, or hard-packed gravel, it reaches its maximum height in canyons west of the Coast Range where there is deep, rich soil and low humidity. It is frequently one of the first migrants to logged or burned forest lands.

In early spring, the young unrolled fronds of this fern, just as they appear above ground, are used as greens and found to be pleasant and nutritious. The brown velvety covering must first be removed; then the shoots are cooked in the manner of asparagus. However, later in the season the fronds are poisonous to stock and probably to human beings as well. In horses and sheep the poison appears to be cumulative, and symptoms may not be evident until some time after the first feedings of the plant.

Except in the case of deficiency in normal pasture plants, animals accustomed to its presence appear rarely to graze fern; or perhaps in some cases they develop an immunity to it. But stock newly introduced to fern-bearing pastures, especially if the grass is low or sparse, may browse it freely, with disastrous results.

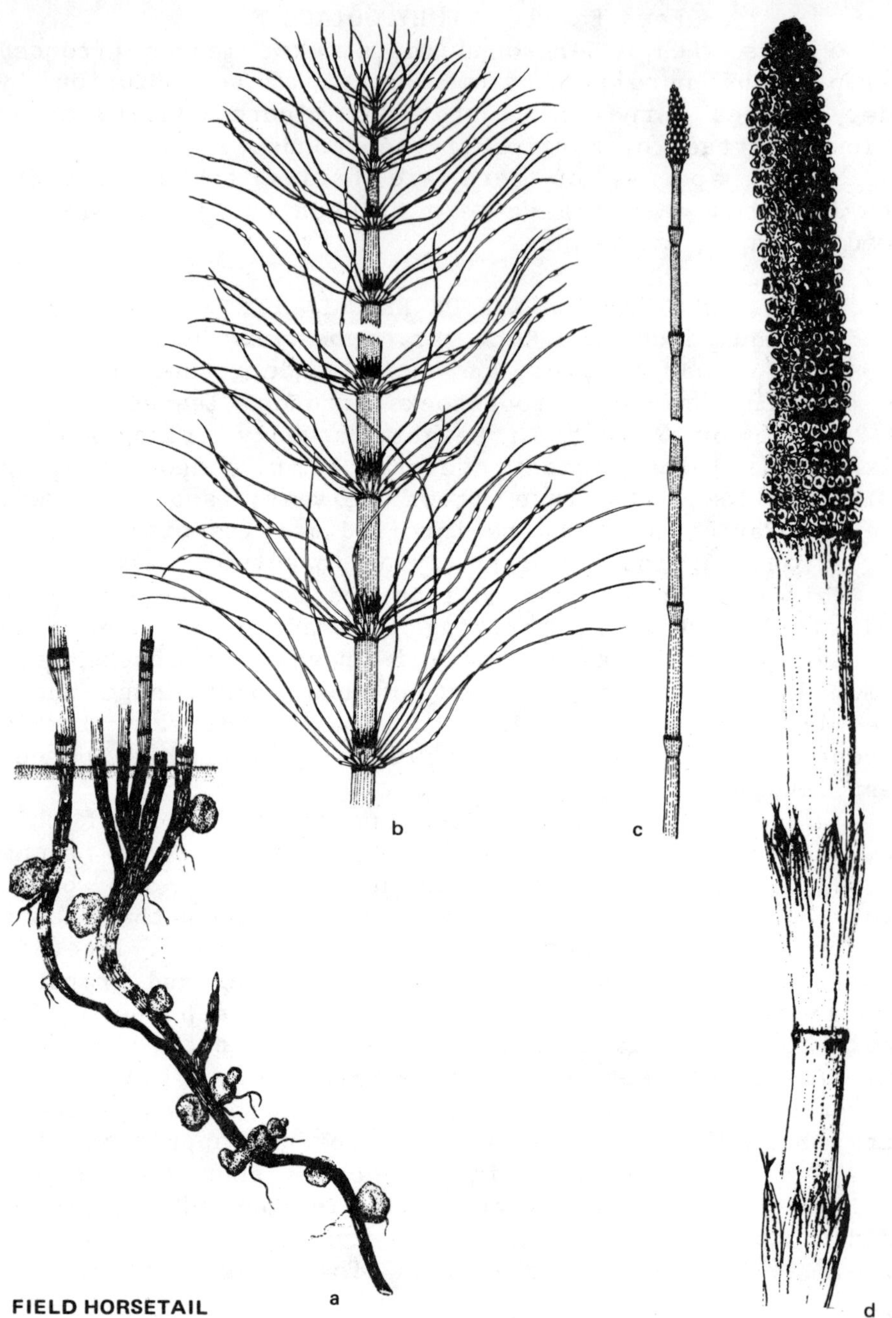

FIELD HORSETAIL
Equisetum arvense
a. tuber bearing rootstock x 1/3
b. sterile shoot x 1/3
c. fertile stem x 1/3

GIANT HORSETAIL
Equisetum telmateia x 4/5

Horsetail Family: EQUISETACEAE

Stems branched or unbranched, ridged and generally roughened, jointed, hollow except at the nodes; leaves minute, scale-like, borne in whorls; spores produced in terminal cone-like structures. In some species the stems above ground die down each year, while in others they may remain for several years. The roughening of the surface of the stems is caused by minute particles of transparent silica. The common name, Scouring Rush, arose from the use of these plants by the pioneers as a scouring agent for pots and pans.

FIELD HORSETAIL

Equisetum arvense L.

Aerial stems annual, plant perennial underground by dark creeping brown-woolly tuber-bearing rootstocks; cone bearing fertile stem arising in early spring, flesh color, erect, 1/2 to 1 foot tall, sheaths thin with generally 8 to 12 dark purplish of brownish teeth; cone 3/4 to 1 1/2 inches long, narrowly ovoid, long-stalked; vegetative stems arising after the fertile stems, more or less erect, 1/2 to 2 feet tall, with many whorls of slender, green jointed branches. The fertile stems die down shortly after the pale green spores are shed, but the vegetative stems die down in the autumn.

This cosmopolitan species grows in moist areas or where the water table is high. Its tendency to spread rapidly underground and its frequent abundance in cultivated overflow land and in low pastures render it an expensive nuisance at times. Its tenacity is remarkable, as exemplified by its re-emergence after being covered by several feet of earth in fills and dikes.

A number of species of Horsetail are known to be poisonous to stock; Field Horsetail, since it is the most common, being the chief offender. It is particularly injurious to sheep and horses and is more likely to cause trouble when fed in hay than in the green state.

GIANT HORSETAIL

Equisetum telmateia Ehrh. var. *braunii* (Milde) Milde

Resembles *E. arvense*, but is much more robust; the sterile stems are usually over 1 1/2 feet tall and the fertile stems 10 to 24 inches with cones 1 1/2 to 4 inches long.

Common in moist low ground from southern Alaska to California, west of the Cascades in our area.

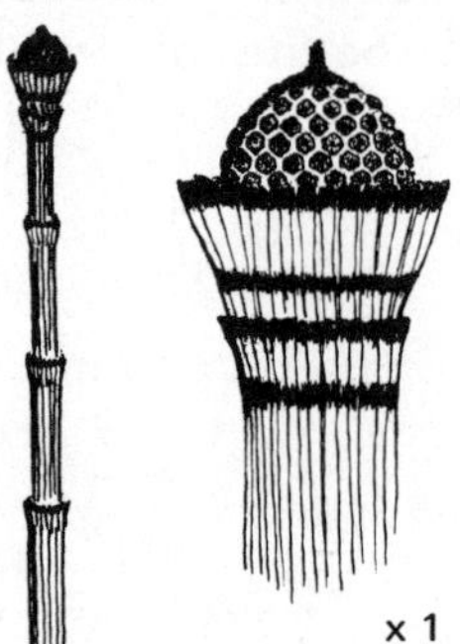

x 1 1/3

x 1/12

SCOURING RUSH

Equisetum hyemale L.

Stems evergreen, all alike and usually unbranched, prominently black-banded at the nodes; cones 3/8 to 1 inch long, abruptly pointed.

Wet ground, especially along streams; throughout our area.

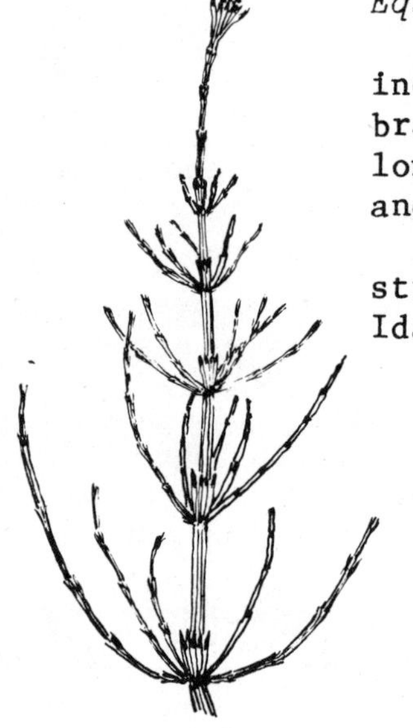

x 1/4

MARSH HORSETAIL

Equisetum palustre L.

Stems annual, nearly all alike, 8 inches to 3 feet tall, the whorled branches varying from few to many; the long-stalked cones 3/8 to 1 inch long and blunt.

Marshes, wet meadows and along streams; Alaska to Washington and Idaho; uncommon in Oregon.

Cattail Family: TYPHACEAE

Herbs with long leaves sheathing at the base, and with erect stems bearing pistillate (seed-producing) flowers in a dense velvety cylinder; the staminate (pollen-producing) flowers, more loosely arranged, borne above the pistillate, and eventually shed.

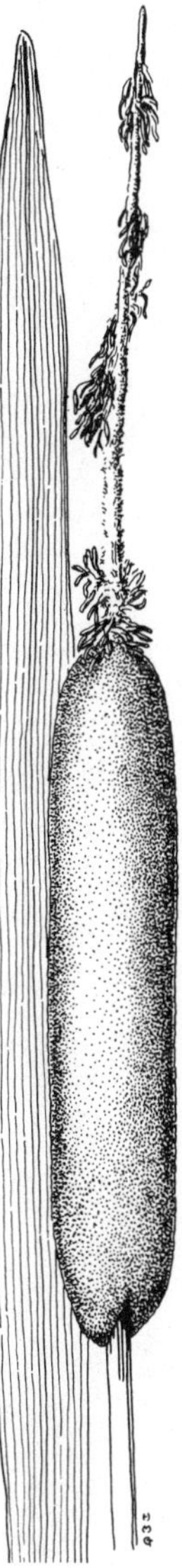

COMMON CATTAIL

Typha latifolia L.

Perennial plants spreading underground by large creeping rootstocks; stems 4 to 9 feet tall, bearing long broad parallel-veined leaves and stout flowering stalks ending in long thick brown spikes; spikes composed at first of two parts, the lower very dense portion consisting of the pistillate flowers, the upper portion bearing the pollen-producing stamens; after pollination the pollen-bearing flowers withering and gradually disappearing, the pistillate portion of the spike thickening and eventually disintegrating to allow the escape of clouds of minute fruits, each equipped with several hair-like bristles to facilitate floating in the air.

Common Cattail, which is known over a large part of the world, is widely distributed throughout our territory. It grows in marshes, ditches, and along margins of streams, sloughs, and lakes. Its ability to creep under ground and thus to spread rapidly makes it a problem in irrigation ditches which may become choked by it in a short space of time; except in irrigated areas, or where it encroaches on watering places, its presence is generally of no material significance. Indians dried the rootstocks and ground them into meal; the young shoots and flowering spikes were used, raw or roasted, as food; and the pollen has sometimes been utilized as a flour substitute.

x 1/3

EGERIA
Elodea densa x 1

Frog's-bit Family: HYDROCHARITACEAE

Water plants, ours submerged, with leaves borne in 2's, 3's or 4's (rarely 6's), and with pollen-bearing and seed-bearing flowers on different plants.

Many species of aquatic plants occur in the Northwest; but since these form an extensive field of study in themselves, and since their importance as weeds is somewhat specialized, few of them will be considered here.

EGERIA or SOUTH AMERICAN WATERWEED

Elodea densa (Planch.) Casp.
(=*Egeria densa* Planch.)

Coarse perennial plant rooting in the mud of ponds and lakes; main body of the plant wholly submerged in water; the stems long and weak, forming branching tangled masses; leaves averaging an inch long or more; borne generally in 4's, occasionally in 6's, dark green; flowers white, 3-petaled, 1/2 to 1 inch in diameter, raised to the water surface by a long slender stalk-like tube.

This species is a native of South America, probably Argentina, but has long been used in the United States in water gardens and aquaria. It escapes from cultivation and becomes a pest in quiet water. This in certain lakes of western Oregon, resulting in such dense growth that fishing and boating become impossible. Since the plants root in the mud on the bottom, only comparatively shallow bodies of water (under approximately 20 feet in depth) are rendered completely unusable by the weed.

It is of the utmost importance that transfer of plants from infested to non-infested lakes be prevented. To date, only pollen-bearing plants have been found in Oregon; therefore propagation in this state appears never to be by seed but only by vegetative means. Portions of broken stems readily take root and become new plants. Pieces of stems may be transported from lake to lake with the movement of boats and outboard motors from one area to another.

A native species, *Elodea canadensis* Richard in Michx. (ELODEA) is found throughout our range, but is much less aggressive. It differs from *E. densa* in its leaves averaging less than 3/4 of an inch long, borne in pairs below, and mostly in 3's above.

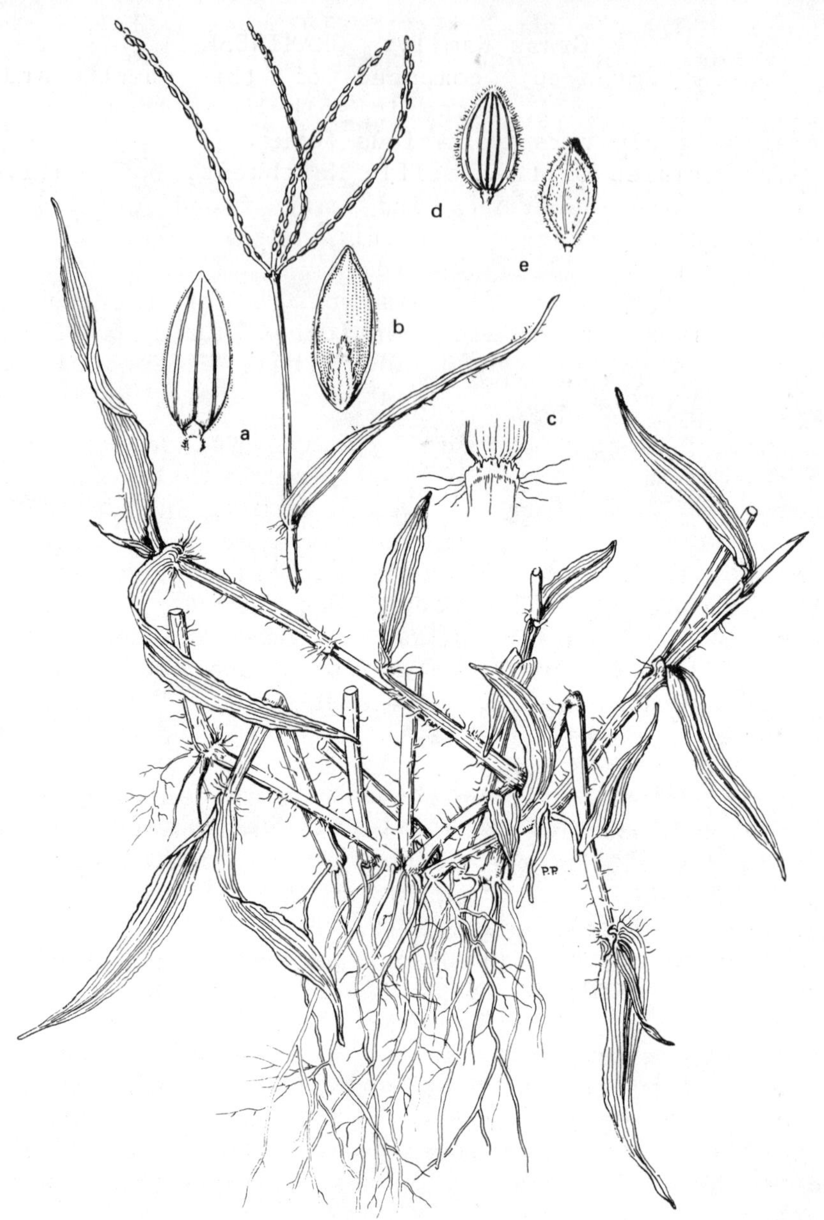

LARGE CRABGRASS
Digitaria sanguinalis x 2/3
a. dorsal view of spikelet x 10
b. ventral view x 10
c. ligule x 2

SMOOTH CRABGRASS
Digitaria ischaemum
d. dorsal view of spikelet x 10
e. ventral view x 10

Grass Family: GRAMINEAE

Leaves 2-ranked, composed of the sheath and the blade; florets small, borne in spikelets, these in spikes or branching clusters of various forms.

The grasses of the Pacific Northwest, both native and introduced are numerous, and many found growing along roadsides or encroaching on cultivated fields may perhaps be designated as "weedy". Inclusion of all, however, is not within the scope of this work. For the most part, those grasses which are considered are restricted to species (mostly introduced) which have escaped all bounds and replaced more valuable pasturage on the range, or have proved difficult to control in gardens and cultivated fields.

The alternate name for this family is the POACEAE.

LARGE CRABGRASS

Digitaria sanguinalis (L.) Scop.

Annual; stems much branched at the base, creeping and rooting at the enlarged nodes (joints), the stems sometimes reaching a horizontal length of several feet, sending up erect flower stalks at intervals; lower leaf sheaths stiff-hairy, leaf blades pale green, 1/8 to 3/8 inch wide, hairy or roughened; flower spikes slender, several pairs generally arising together from the tip of the stalk, and often one or more attached below.

A native of Europe, this grass has now spread over most of the United States, and is a noxious weed in lawns and gardens. Although only an annual, its stolon-like method of progress over the ground gives it an advantage over its cultivated competitors.

SMOOTH CRABGRASS

Digitaria ischaemum (Schreb.) Schreb. ex Muhl.

Annual. This species considerably resembles the preceding, but is shorter, finer, and with smooth leaf sheaths. It, too, has been introduced from Europe, and is becoming alarmingly well established along roadsides and as a lawn weed in parts of the Pacific Northwest.

Both species here described carry on their heaviest growth in summer, drop their seeds, and die with the beginning of cold weather, leaving bare spots in an infested lawn. Their frequent reddish color, also, makes them undesirable on a green turf.

BERMUDA GRASS
Cynodon dactylon x 2/3
a. spikelet x 10
b. ligule x 4

BERMUDA GRASS

Cynodon dactylon (L.) Pers.

Perennial with long scaly rootstocks; stems more or less creeping and freely rooting at the nodes; the flowering stems upright, 3 to 16 inches tall, each bearing at the apex a group of 4 or 5 (rarely 7 to 9) slender spikes 1 to 2 inches long; axis of each spike flattened, bearing on one side two rows of spikelets; seed contained in a hardened, shining covering.

This Eurasian species is widespread in most of the warmer parts of the world. As early as 1807 it was listed as one of the most important grasses of the Southern States, and it now extends into many northern localities. In our area it is reported from scattered localities both east and west of the Cascades. In the South it is valuable for hay, pasture, and as a lawn grass, but certain of the qualities which give it value, such as its resistance to drought, tolerance to alkali, vegetative propagation and abundant seed production, render it a pest in the North Pacific States, where it crowds out plants better suited to our needs.

RATTAIL FESCUE

Festuca myuros L.
(=*Vulpia myuros* (L.) C. C. Gmelin)

Slender smooth annual; stems 4 to 20 inches tall; leaf blades folded, less than 1/16 of an inch wide; spikelets 3- to 6-flowered, rarely more, borne in narrow panicles; florets with awns 1/4 to 5/8 of an inch long.

Native of Europe; weedy in dry wasteland and overgrazed areas.

x 2

SIXWEEKS FESCUE

Festuca octoflora Walt.
(=*Vulpia octoflora* (Walt.) Rydb.)

Similar to *Festuca myuros* but with spikelets 7- to 15-flowered and florets with a awn 1/4 inch long or less.

Common in dry open areas throughout much of North America.

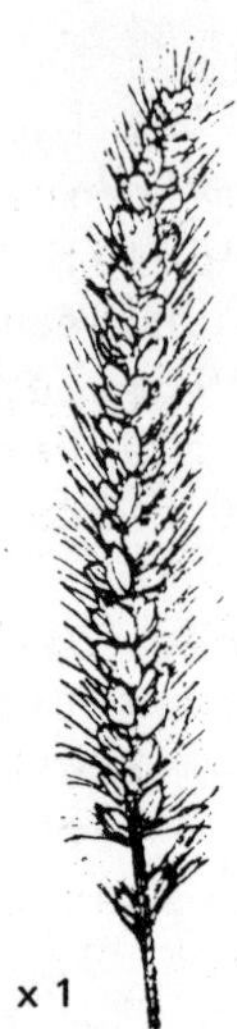

x 1

YELLOW FOXTAIL or BRISTLEGRASS

Setaria glauca (L.) Beauv.
(=*Setaria lutescens* (Weigel) Hubb.)

Annual, usually much branched from the base and often rooting at the lower nodes; stems up to 3 feet long; leaf blades 1/8 to 1/2 of an inch wide; spikelets borne in a spike-like panicle; spikelets 2-flowered, each spikelet subtended by 5 to 10 stiff bristles.

Native of Europe; found occasionally on both sides of the Cascades in moist waste places, ditchbanks and in irrigated cultivated land.

GREEN FOXTAIL or BRISTLEGRASS

Setaria viridis (L.) Beauv.

Similar to *Setaria glauca* but with the spikelets each subtended by 1 to 3 or rarely 4 stiff bristles.

Introduced from Eurasia; now a widespread weed in moist areas, particularly troublesome in gardens and irrigated fields.

x 1 x 4

MEADOW FOXTAIL

Alopecurus pratensis L.

Stout perennial; stems 1 to 3 feet tall, sometimes rooting at the lower nodes; leaf blades less than 1/2 of an inch wide; spikes 1 1/4 to 4 inches long; spikelets crowded, each spikelet with 1 floret which has a slender, abruptly bent awn.

This European species is found in wet, even swampy ground, throughout much of North America.

GERMAN or CREEPING VELVETGRASS

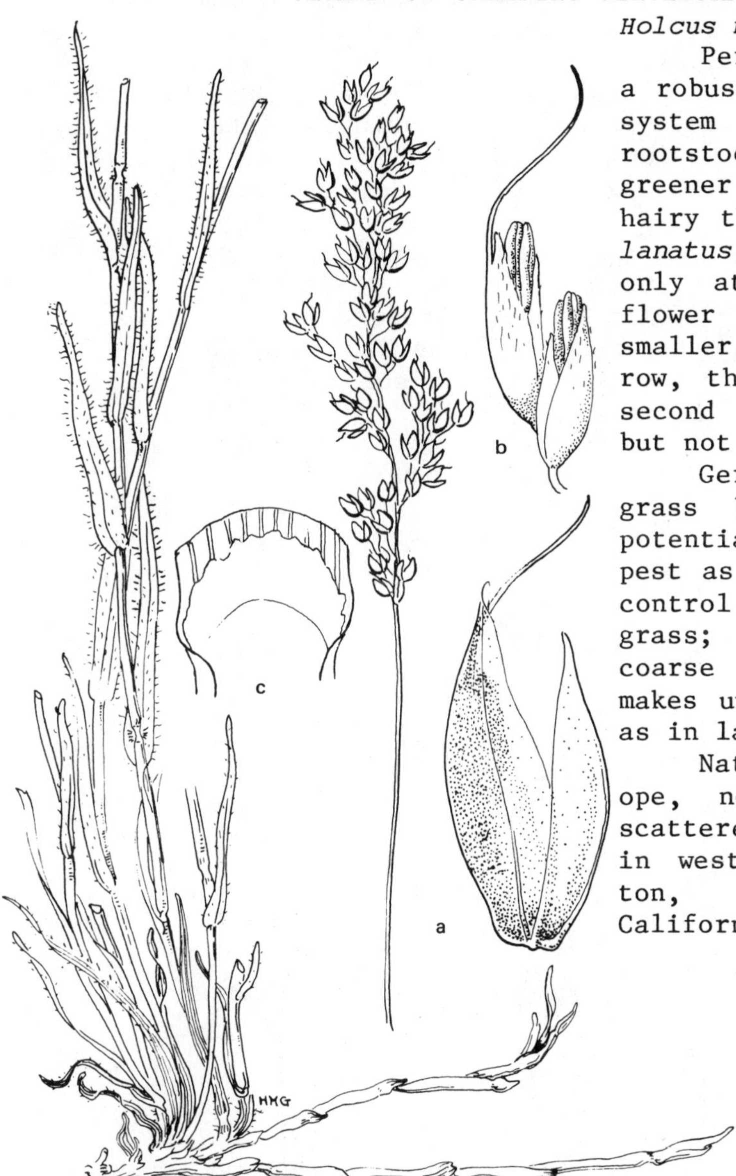

GERMAN VELVETGRASS
Holcus mollis x 2/3
a. spikelet x 10
b. florets x 10
c. ligule x 4

Holcus mollis L.

Perennial, with a robust underground system of spreading rootstocks; plant greener and less hairy than in *Holcus lanatus*; stems hairy only at the nodes; flower panicles smaller, more narrow, the awn of the second floret bent but not hooked.

German Velvetgrass has all the potentialities of a pest as difficult to control as Quackgrass; and its coarse low growth makes unsightly areas in lawns.

Native of Europe, now found in scattered localities in western Washington, Oregon and California.

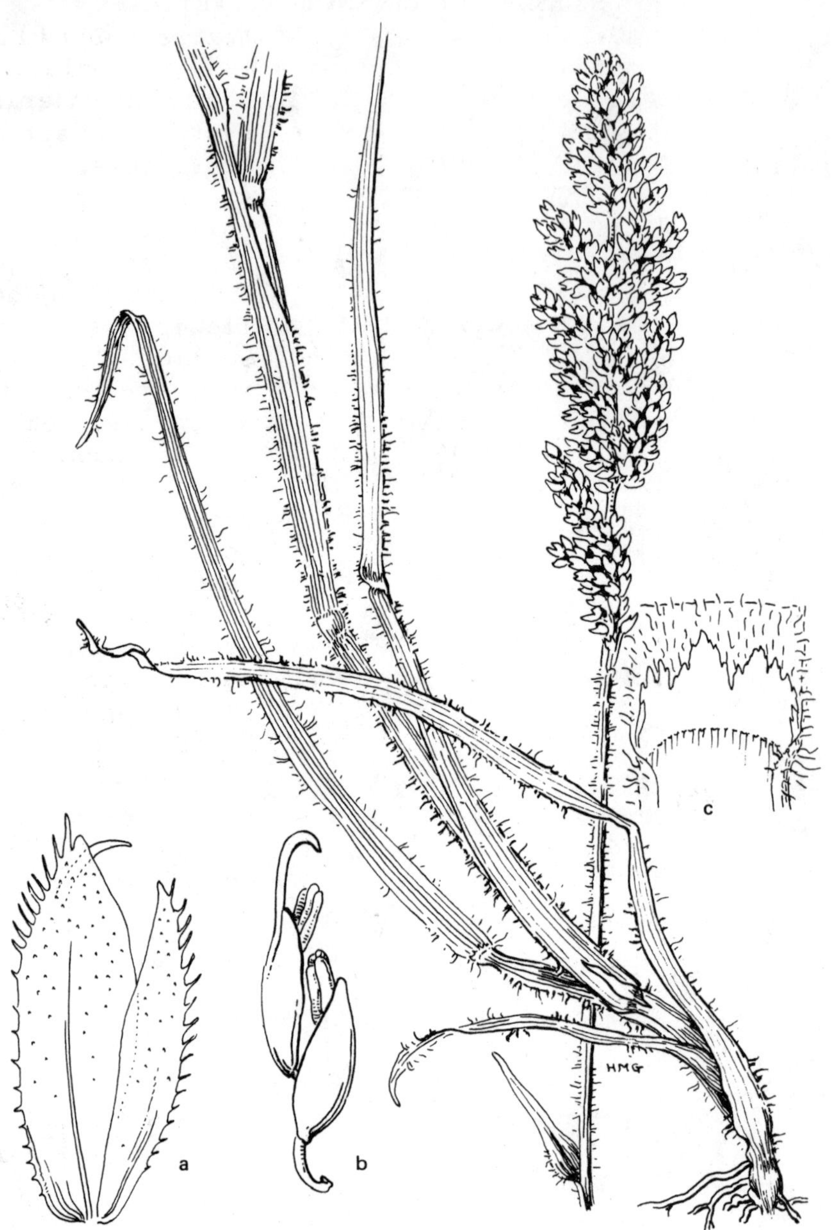

VELVETGRASS
Holcus lanatus x 2/3
a. spikelet x 10
b. florets x 10
c. ligule x 4

VELVETGRASS

Holcus lanatus L.

Perennial; stems tufted, 1 to 3 or more feet tall, the entire plant more or less grayish velvety; leaves broad, long-pointed at the apex; flower panicles plume-like, dense, pale-green to purplish, hairy, 3 to 6 inches long; the second floret of each spikelet possessing a microscopic curved, hook-like awn.

A European introduction, this grass has become very abundant west of the Cascades and is somewhat often found east of the mountains as well. It frequently invades lawns, replacing the normal turf with patches of coarse stems and foliage, but is also common in disturbed sites and in open woods.

MEDUSAHEAD

Taeniatherum caput-medusae (L.) Nevski
(=*Elymus caput-medusae* L.)

It is possible that our material should be called *Taeniatherum asperum* (Simonkai) Nevski.

Slender annual 6 to 24 inches tall; leaf blades more or less rolled, less than 1/8 of an inch wide; spikelets in a short dense bristly spike; spikelets usually 2 per node, densely crowded; bristles 1 to 4 inches long.

This Eurasian weed is occasional in eastern Washington in sagebrush areas, but more abundant in Oregon, California and Idaho. In Oregon it is found on both sides of the Cascades.

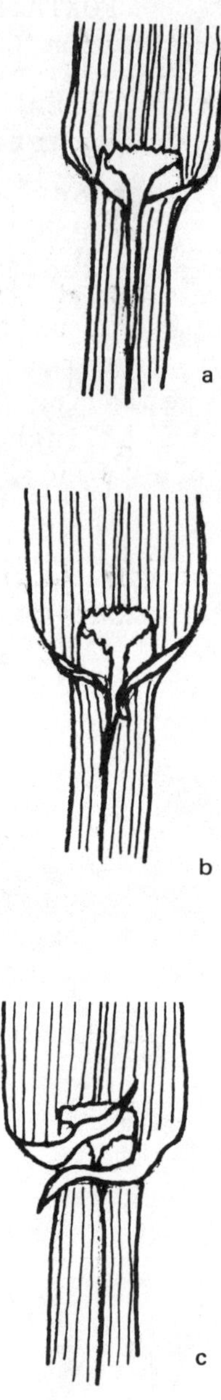

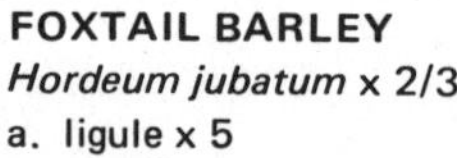

FOXTAIL BARLEY
Hordeum jubatum x 2/3
a. ligule x 5

MEDITERRANEAN BARLEY
Hordeum geniculatum
b. ligule and auricle x 5

WILD BARLEY
Hordeum leporinum
c. ligule and auricles x 5

FOXTAIL BARLEY or SQUIRRELTAIL BARLEY

Hordeum jubatum L.

Perennial; stems tufted, erect, or angled at the lower joints, 1 to 2 feet tall; leaf blades 1/8 to 1/4 inch wide, the sheaths smooth to densely hairy; spike pale green, more or less nodding, 2 to 4 inches long, the awns slender, spreading, 3/4 to 2 1/2 inches long, pale green or slightly reddish tinged.

This grass, which is a native of North America, is palatable in its young stages to grazing animals but becomes troublesome with maturity of the needle-like awns. If grazed or present in hay at this later stage, the awns can inflict much damage by penetrating the delicate tissues of mouth and nose and resulting in abscesses of these areas, or even reaching the stomach with similar after-effects.

The seed spikes readily break apart, and may be carried by the hair and wool of animals to new areas. The plant is a heavy seeder, and once established is difficult to control. It may become a pest in hay and grain fields, particularly east of the Cascades.

MEDITERRANEAN BARLEY

Hordeum geniculatum Allioni
(=*Hordeum hystrix* Roth)

Smooth to densely hairy annual; stems 4-16 inches tall; isolated plants tend to be much branched and spread out at the base, while crowded plants grow erect; leaf blades 1/16 to 3/16 of an inch wide; some of the leaves with at least one small claw at the base of the blade; spikes erect, 1/2 to 1 1/2 inches long; awns to 1/2 inch long; seed spikes readily breaking apart.

This weedy European species is common on both sides of the Cascades in moist, even alkaline, or dry waste areas.

WILD BARLEY

Hordeum leporinum Link

Similar to Mediterranean Barley but with a pair of small claws at the base of the blade on most leaves and with spikes 1/2 to 4 inches long.

Native of Europe; now common in waste places throughout our area.

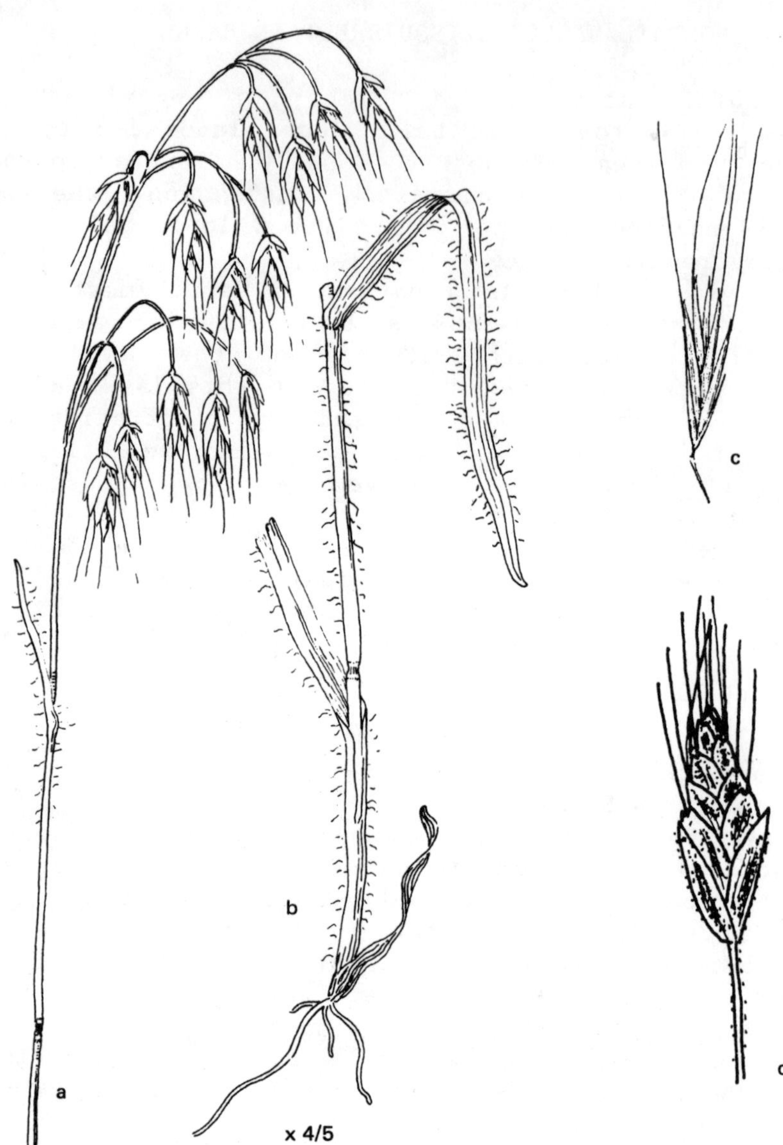

DOWNY BROME
Bromus tectorum
a. inflorescence x 4/5
b. stem x 4/5

RIPGUT BROME
Bromus rigidus
c. spikelet x 1/2

SOFT CHESS
Bromus mollis
d. spikelet x 2

DOWNY BROME or CHEATGRASS BROME

Bromus tectorum L.

Annual, 3/4 to 3 feet tall; leaf sheaths and blades soft-hairy; "heads" loose, drooping, often purplish; spikelets long-bearded; plant readily distributed by seed.

This is an introduced species that reached us from Europe. Its adaptability to unfavorable conditions, such as exist on overgrazed lands, has aided and speeded its naturalization, until it is now one of our most abundant and widespread plant species. Where native grasses have been destroyed by misuse of grazing lands, Downy Brome has taken over a large acreage, providing in most cases less forage value than that possessed by the original grass inhabitants. And, being an annual and thus more dependent upon weather conditions, its fluctuation in forage value is far greater from year to year than that of native perennials.

In a favorable season when germination occurs in the fall and good growth is attained in the spring, Downy Brome provides a certain amount of early pasturage. But, as the plant matures the long stiff beards may cause serious damage to mouths and eyes, even resulting in severe infections or death.

This species is common in much of the western United States. It is widespread east of the Cascades; west of the Cascades this grass is found on roadsides and waste land, and can adapt itself to gravel banks and stony hillsides normally unfavorable to most other plants.

RIPGUT BROME

Bromus rigidus Roth

Similar to *Bromus tectorum* but with much larger spikelets. Spikelets 1 to 1 1/2 inches long with awns as much as 2 1/4 inches long.

This species is widely introduced, but is more abundant west of the Cascades.

SOFT CHESS

Bromus mollis L.

Annual, 8 inches to 3 feet tall, usually "soft hairy" throughout; blades about 1/8 of an inch in width; "heads" erect, the spikelets more or less crowded; spikelets 3/8 to 7/8 of an inch long, 5- to 7-flowered, the awns 1/4 to 1/2 inch long.

A native of Europe, this species is found throughout the Pacific states in dry waste places.

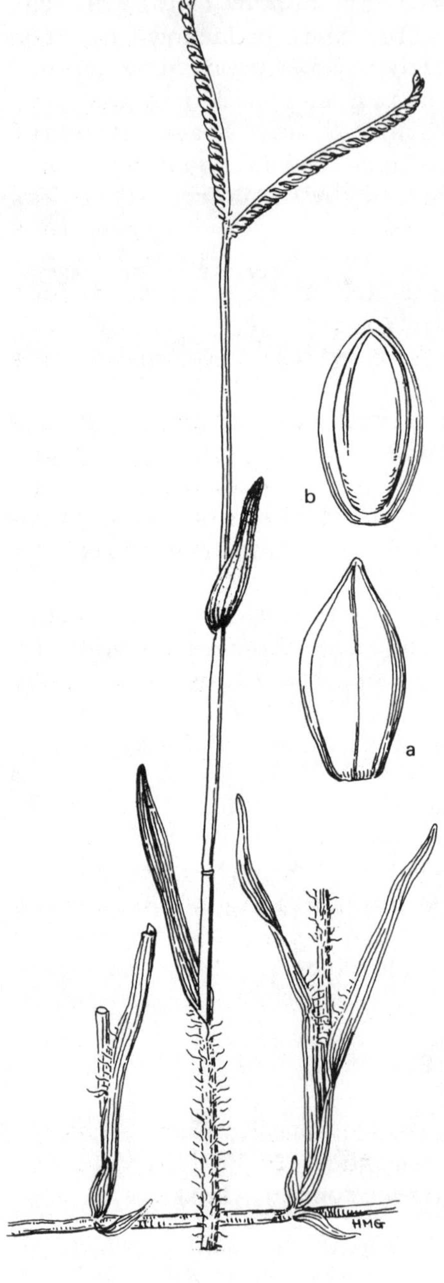

KNOTGRASS

Paspalum distichum x 2/3

a. dorsal view of spikelet x 10

b. ventral view x 10

KNOTGRASS

Paspalum distichum L.

Perennial by stout creeping rootstocks, these sending up stems 1/2 to 2 feet tall; stems bent at the lower joints, branching, with joints enlarged; leaves flat, often short, tapering to a point, smooth or sometimes long-hairy on the upper surface and at base of the blade; spikes slender, generally borne in a pair at the summit of the stem, or occasionally with an extra spike below; spikelets borne in a double row on one side of the axis.

Knotgrass is a native of this continent, probably originating in the American tropics. It thrives in moist situations along streams and in lowland pastures and fields. With its ability to creep rapidly over an area, establishing itself as it goes, it can crowd out better plants and thus become a pest. It is a special menace to irrigation ditches, where its rapid rank growth may succeed in choking the flow of water. In our area it is most common west of the Cascades and along the Columbia river.

PERENNIAL RYEGRASS

Lolium perenne L.

Tufted perennial; stems 1 to 3 1/2 feet tall; leaf blades less than 1/4 of an inch wide, usually bearing a pair of short claws at the base; spikelets sessile and solitary at the nodes, 3- to 12-flowered, florets without awns.

Widely cultivated for use as forage, cover crops and in lawns; escaped and becoming weedy in waste ground, especially along roadsides.

Lolium perenne and *Lolium multiflorum* cross freely.

x 1/2

ITALIAN RYEGRASS

Lolium multiflorum Lam.

Annual, biennial, or short-lived perennial; stems 1 1/2 to 4 feet tall; leaf blades usually flat, 1/8 to 3/8 of an inch wide, rough on the upper surface, most leaves with a pair of small claws at the base of the blade; spikelets sessile and solitary at each node, 5- to 15-flowered, florets with slender awns up to 1/4 of an inch long.

This European species is commonly cultivated and is escaped and weedy along roadsides and in waste places throughout much of North America.

x 3/4

DARNEL

Lolium temulentum L.

Annual; stems 1 1/2 to 3 feet tall; leaf blades flat, 1/8 to 3/8 of an inch wide, the pair of claws at the base of the blade very prominent; spikelets usually 5- to 8-flowered, the florets with or without an awn.

Native of Europe; only occasional as a weed in our area.

x 1

TUFTED HARDGRASS

Sclerochloa dura (L.) Beauv.

Low annual with generally many erect or spreading flower stalks 1 to 5 inches long; leaf sheaths closed below, conspicuously veined; leaf blades generally flat, rarely folded; flower clusters dense, somewhat one-sided; often more or less enclosed in the upper leaves; scales of the florets prominently veined.

This is a European grass which is well established in eastern Oregon and Washington and in southwestern Idaho as a lawn pest. Its scant rather gray-green foliage and large flower clusters on short stems make it conspicuous and unattractive in a green lawn. It spreads by seeds only, but since an abundance of these is produced, it may quickly invade a large area.

TUFTED HARDGRASS

Sclerochloa dura x 1

a. spikelet x 3

b. tip of leaf x 5

c-e. variations in ligule x 3

ANNUAL BLUEGRASS

Poa annua L.

Annual; stems somewhat flattened, tufted, spreading or erect, 2 inches to 1 foot long, sometimes forming dense clumps; leaves bright green, soft; flower clusters pyramid-shaped, open, with spreading branches.

This is one of the first and, in spite of its dwarf size, one of the most conspicuous grasses to begin growth in very early spring or in sunny days of winter. The brilliant green of its foliage and its lush growth would seem to insure its value as a lawn grass. However, being an annual and of very early growth, it often completes its life cycle before other grasses have scarcely begun to grow; and its early flowering and subsequent death result in unsightly patches in the lawn.

It is one of the earliest, too, of garden weeds, and under some conditions may be useful as a ground cover in winter and early spring. Its early ripening, however, insures a heavy crop of seed which, in the sprinkled or irrigated garden, may give trouble all summer.

Annual Bluegrass is a native of Europe; it is nearly cosmopolitan and in our area is more common west of the Cascades.

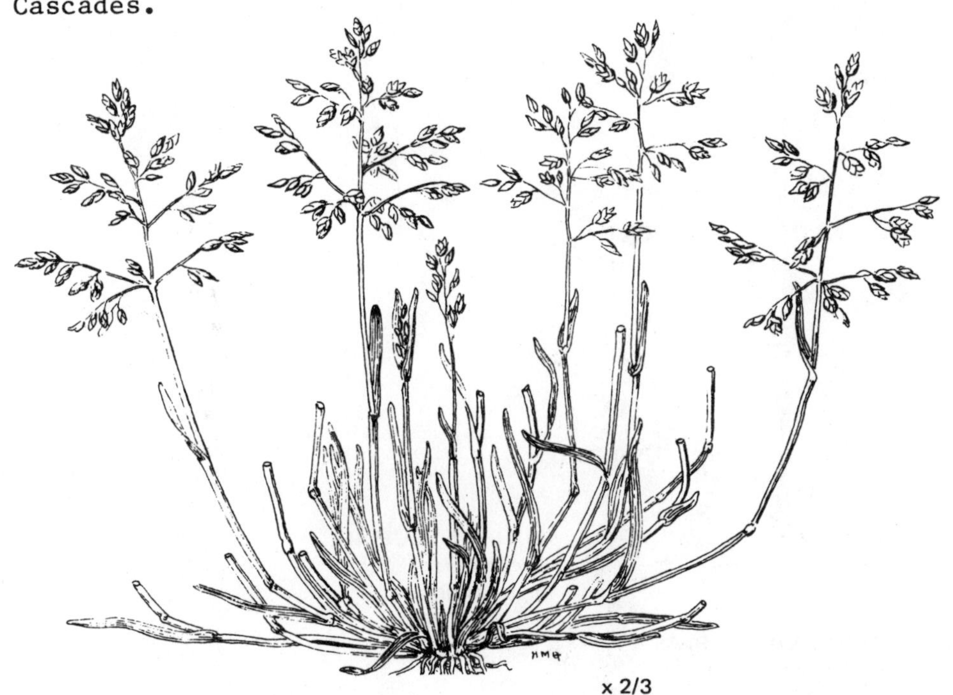

x 2/3

c

CANADA BLUEGRASS
a. *Poa compressa* x 2/3

BULBOUS BLUEGRASS
Poa bulbosa
b. inflorescence x 1
c. bulbil x 4

CANADA BLUEGRASS

Poa compressa L.

Perennial by stout creeping rootstocks; stems often curved at base, erect above, flattened, slender, 1/3 to 2 feet or more tall, bluish-green; leaf blades short, 1/8 inch wide; flower cluster narrow, 1 to 3 or 4 inches long, often crowded.

This bluegrass, which, like Canada Thistle, a native of the Old World rather than Canada, was perhaps introduced as a pasture grass on poor soil. Its ability to thrive in such conditions has made it of value for a forage crop where no better plants will grow. But its ability to spread underground and crowd out other plants has made it a pest in cultivated fields, gardens, and lawns, throughout much of North America.

It somewhat resembles Kentucky Bluegrass, but is typically shorter, with shorter stiffer leaves, narrower and more compact flower cluster, and a distinctly flattened stem. The latter can readily be detected by a lens, or by rolling the stem through the fingers.

Canada Bluegrass normally spreads more rapidly underground than Kentucky Bluegrass, and is difficult to control in gardens, since cultivation results in segmenting and scattering the rootstocks which still retain their ability to produce new plants. The flowering stems are not tufted, as in Kentucky Bluegrass, but generally arise singly from various points of the rootstock and, if undisturbed, eventually form a colony over a wide area.

BULBOUS BLUEGRASS

Poa bulbosa L.

Smooth tufted perennial, 1/2 to 2 feet tall, purplish at least at the base; flowers usually modified into purplish bulbils, these up to 3/4 of an inch long.

This European species is weedy in pastures and various disturbed sites, mainly east of the Cascades.

Additional species of *Poa* may tend to be weedy in certain situations.

REED CANARYGRASS
Phalaris arundinacea
a. stem x 2/3
b. inflorescence x 1
c. floret x 6

d. **COMMON BENTGRASS**
Agrostis tenuis x 1/3

REED CANARYGRASS

Phalaris arundinacea L.

Stout perennial from large rootstocks; stems 2 to 7 feet tall; leaf blades flat, 1/4 to 3/4 of an inch wide; panicle more or less compact at first, then the branches spreading.

Wet ground, along streams and in marshes. This species is extremely aggressive in areas with adequate moisture.

COMMON or COLONIAL BENTGRASS

Agrostis tenuis Sibth.

Perennial from short rootstocks and/or stems creeping and rooting along the surface of the ground; stems less than 2 feet long; leaf blades flat or folded, 1/16 to 3/16 of an inch wide; flower panicles open, the branches thread-like, not spikelet-bearing at the base; spikelets each with a single floret, often purplish-tinged.

This species is a common lawn grass throughout the northern states. It escapes and can be found in moist ground, west, occasionally east, of the Cascades.

BARNYARD GRASS

Echinochloa crus-galli (L.) Beauv.

Annual; stems 1 to 5 feet tall; leaf blades 3/8 to 5/8 of an inch wide; this is the only species of grass in our area without a ligule at the junction of the sheath and the blade; spikelets in panicles; spikelets about 1/8 of an inch long, with or without a slender awn.

Weedy in wet ground, especially in irrigated areas and in ditches.

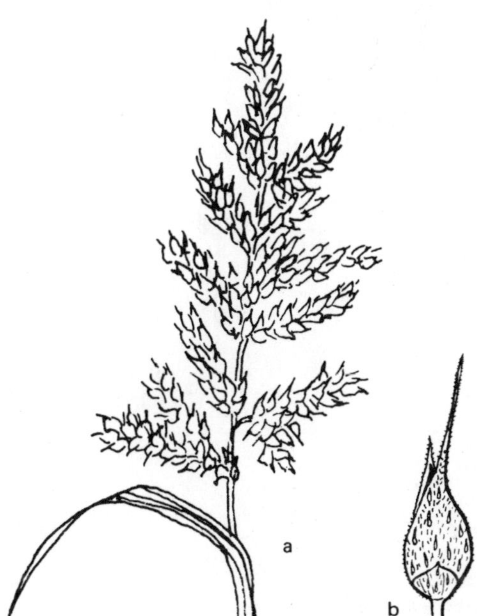

BARNYARD GRASS
Echinochloa crus-galli
a. inflorescence x 3/4
b. spikelet x 2 2/3
c. floret x 2 2/3

JOHNSONGRASS
Sorghum halepense
a. stem x 1/3
b. paired spikelets x 3
c. portion of inflorescence x 2/3

JOHNSONGRASS

Sorghum halepense (L.) Pers.

Perennial by long stout creeping rootstocks; stems 2 to 5 feet tall, stout, branching near the base, swollen at the joints; leaves sometimes reaching 3/4 inch in width, thick, with conspicuous midveins; flower cluster (panicle) 1/2 to 1 foot or more long, open or somewhat dense, green or purplish, the fertile spikelets firm, each at first provided with a bent needle-like awn.

This grass, which is a native of the Mediterranean area but not uncommon in the United States, is widely cultivated and is occasionally reported as a troublesome weed in the Northwest. Its extensive system of rapidly spreading robust rootstocks enables the grass to store enormous quantities of food for use in sending up new plants. Its invasion of cultivated land, under conditions favorable to its development, is serious because its growth habit gives it great advantage over most crops. In such areas it spreads rapidly, not only pre-empting the raw materials needed by the legitimate crop for food manufacture, but also by the extraordinary growth of its rootstocks, literally crowding out or distorting the underground parts of neighboring plants.

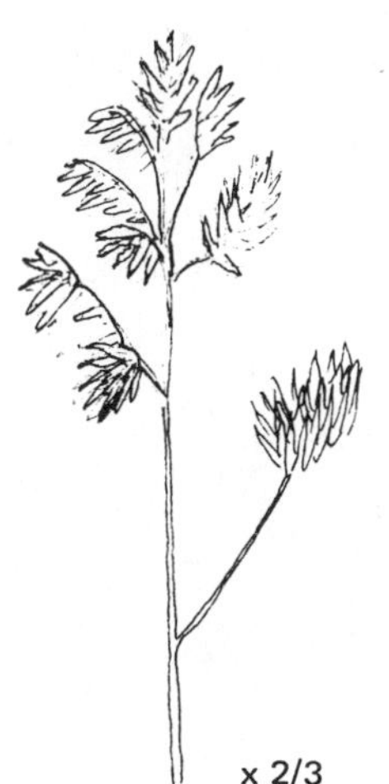

x 2/3

ORCHARDGRASS

Dactylis glomerata L.

Tufted perennial; stems up to 4 feet in height; leaves more or less roughened, sheaths flattened and keeled, blades 1/8 to 1/2 inch wide; flowers borne in 1-sided clusters on stiff panicle branches; florets usually awn-tipped.

This Eurasian species is widely cultivated for pasture and for use as hay. It has, however, commonly escaped and is more or less weedy along roadsides and in disturbed habitats throughout most of the United States.

TUBER OATGRASS
Arrhenatherum elatius var. *bulbosum* x 1/3

TUBER OATGRASS

Arrhenatherum elatius (L.) Presl var. *bulbosum* (Willd.) Spencer

Perennial; bearing, below ground, bulb-like structures in chains resembling strings of beads; stems 2 to 6 feet tall; leaf blades flat, roughened, 3/8 inch or less in width; spike 1/2 to 1 foot long, rather loose, the branches short, whorled, bearing spikelets from base to tip; each spikelet with 2 florets, the lower with a bent awn.

This species, either with or without the bulbous variety here described, is common on the Pacific slope, and occurs sporadically east of the Cascades. The grass is valuable in pastures; but the bulbous variety, which is less frequently encountered, proves troublesome in cultivated ground. The so-called "bulbs," attached to each other in long rows, are easily separated by cultivation, and each is capable of starting a new colony.

WILD OAT

Avena fatua L.

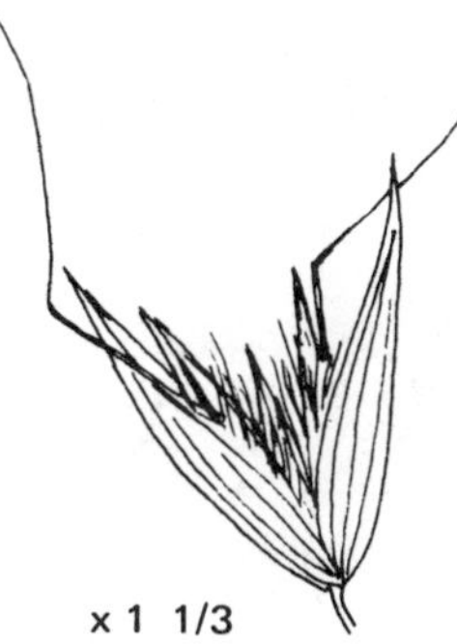
x 1 1/3

Stout annual; stems 1 1/2 to 3 feet tall or more; leaf blades 1/8 to 5/8 of an inch wide, often "soft hairy"; spikelets large, usually with 3 florets, first 2 florets with awns 3/4 to 2 inches long, the awns abruptly bent, the portion below the bend strongly twisted.

A weed of waste ground and cultivated fields.

x 1 1/3

Avena sativa L., the cultivated Oat, occasionally escapes and may be found along roadsides and in fields, but apparently does not persist long out of cultivation. It may be separated from Wild Oat by the awns, which, if present, in cultivated oats occur only on the first floret in each spikelet and are not abruptly bent.

QUACKGRASS
Agropyron repens x 2/3

QUACKGRASS

Agropyron repens (L.) Beauv.

A coarse perennial reproducing both by seed and by creeping rootstocks, the latter often numerous, yellowish, in porous soil sometimes becoming several feet long, branching, and sending up new plants from the nodes or joints; stems above ground erect, generally dark gray-green, from 1 to 4 feet tall; leaves broad, flat, long-pointed at the apex, each leaf with a pair of small claws at the rounded base of the blade; spikelets arranged in 2 rows in a long spike, awnless or with short awns.

Quackgrass, which is native to the Mediterranean area, has spread over a large part of North America except for the dry desert areas. A number of species of *Agropyron* occur in our area, most of them native, some useful, a few weedy, but no other species of this group equals Quackgrass as a pest or in difficulty and expense to control. While it has some value as forage and is sometimes used as a soil binder, its rapid underground distribution in cultivated ground, its ability to crowd out almost any competitor, and its development of a vast storage depot in its system of rootstocks makes it a formidable enemy to crops.

From other species of *Agropyron* found in our region, Quackgrass can be distinguished by its wide flat leaves which frequently bear a few scattered long hairs on the upper surface, by the yellowish rootstocks, and by a peculiarity of the rootstock scales which split first from the base and for a time remain in contact at the tips.

The extraordinary growth pressure expended in the progress of the slender pointed rootstocks through the soil is illustrated by the fact that they often penetrate the roots and tubers of competitive plants in the garden or field. In fact, in crowded compact earth, the rootstocks may be found turning back upon themselves, their sharp tips growing through their own tissues.

Three factors contribute toward making this one of our most noxious weeds of cultivated ground, namely: the ability of its mass of branching underground stems to maintain itself in inaccessible places about the roots of shrubbery and under the margins of walks and roads; the fact that every joint of the rootstock, even when torn from the parent plant, is capable of producing new growth; and the high germination ratio of the seeds.

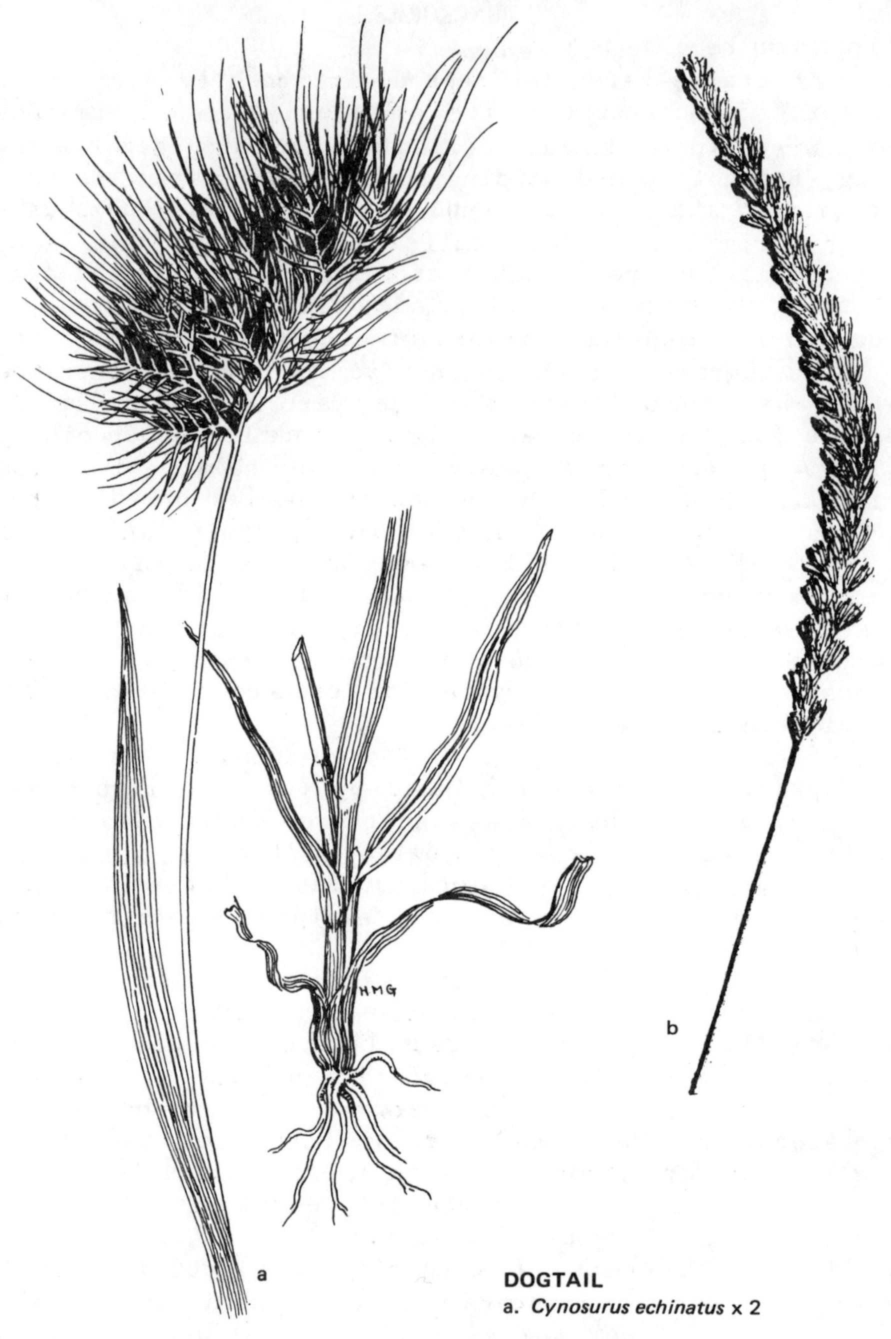

DOGTAIL
a. *Cynosurus echinatus* x 2

CRESTED DOGTAIL GRASS
b. *Cynosurus cristatus* x 1

DOGTAIL

Cynosurus echinatus L.

Annual; stems slender, 1/2 to 1 1/2 feet tall; leaf blades short and early becoming dry; head short, broad, one-sided, bristly, the fertile spikelets intermixed with feather-like sterile ones.

This is an introduced grass which has spread widely over open hills west of the Cascades and is particularly abundant in the Willamette Valley. The leaves are few, small, and so early withered that forage value of the grass is inconsequential; while the heads, which are dry and bristly at maturity, are not relished by stock. The plants are readily adapted to thin soil and rocky areas, crowding out whatever pasturage of value may originally have grown there. The stiff bristles are disagreeable to pedestrians, and dislodged heads or portions of heads cling to clothing of man and to wool and fur of animals.

Through the summer this grass carpets, with a completely useless tinder-dry covering, the hills which have succumbed to its invasion, and may prove to be as serious a fire hazard as Downy Brome or Foxtail Barley.

A perennial species, *Cynosurus cristatus* L. (CRESTED DOGTAIL GRASS) is occasionally cultivated and sometimes becomes weedy west of the Cascades and in Idaho.

WITCHGRASS

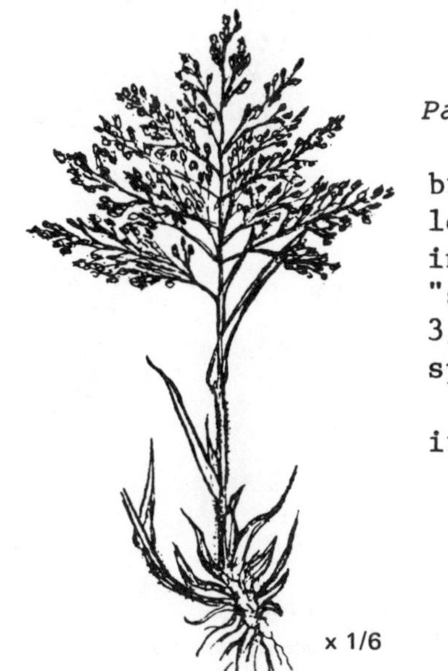
x 1/6

Panicum capillare L.

Annual; stems usually much branched from the base and more or less spreading to nearly erect, 6 inches to 2 feet tall; usually "soft-hairy" throughout; leaf blades 3/16 to 9/16 of an inch wide; spikelets in large open panicles.

Weedy in moist ground and irrigated land.

SANDBUR
Cenchrus longispinus x 2/3
a. ligule x 4
b. spikelet x 4

SANDBUR

Cenchrus longispinus (Hack.) Fern.

Annual, or sometimes behaving as a perennial; stems tufted, spreading horizontally, often branching and forming mats from a few inches to several feet in diameter; leaf sheaths enlarged, flattened; seeds borne in a short spike of spiny "burs," the spines stiff, sharp, and spreading, the burs generally densely woolly at base of spines.

This species of sandbur, unlike the majority of our weeds, is a native of North America; but its original home apparently was subtropical areas from which it has spread northward. It has long been known in restricted locations east of the Cascades, but is now found in a few localities west of the Cascades. It is spreading rapidly along the sandy shores of the Columbia and Snake rivers.

Where abundant the plant is a pest in orchards, vineyards, on the banks of irrigation ditches, and in cultivated fields. It is dangerous to sheep on range lands, the spines injuring the mouths, noses, and eyes of the animals, with infected sores often resulting. To man, also, the burs are irritating. They may be carried in clothing and in the wool and fur of animals, frequently thus being widely distributed.

SILVER HAIRGRASS

Aira caryophyllea L.

Delicate annual, 2 to 11 inches tall; leaf blades less than 1/16 of an inch wide; flower panicle open; spikelets with 2 florets, each floret with a thread-like awn.

West of the Cascades in gravelly soil.

YELLOW NUTGRASS
Cyperus esculentus x 1/2

Sedge Family: CYPERACEAE

Stems often 3-angled; leaves 3-ranked about the stem, but sometimes reduced to sheaths only; sheaths continuous, or, in age, sometimes splitting raggedly on one side.

YELLOW NUTSEDGE

Cyperus esculentus L.

Perennial by creeping slender rootstocks producing small hard tubers; stem stout, 1 to 3 feet tall, 3-angled, bearing long leaves from near the base and a whorl of leaf-like bracts associated with the flower cluster; lower leaves sometimes longer than the stem; bracts at the bases of stalked flowering branches as long as or longer than these branches; branchlets of each flowering branch slender, 1/4 to 3/4 inch long, each bearing 2 rows of minute florets.

Although this species is now common in America, it evidently originated in the Old World. It grows in moist places and is of consequence in the Pacific Northwest because of its frequent occurrence in cultivated land. It reproduces principally by the small fleshy hard tubers developed on the rootstocks, and these are not only difficult to control but may inadvertently be confused with small lily bulbs.

Cyperus strigosus L. (False Nutsedge) occurs in similar habitats and is sometimes confused with Yellow Nutsedge, to which it is closely related. The absence of rootstocks and tubers, however, distinguishes it from that species and also renders it of little importance as a weed.

Another species of *Cyperus* is encountered in southwestern Oregon and occasionally in the Willamette Valley. *Cyperus eragrostis* Lam. (Tall Umbrellaplant) differs from *Cyperus esculentus* in that its rootstocks lack tubers and it has closely congested flower clusters. It occurs along waterways and is sometimes locally abundant in roadside ditches.

Other members of the Cyperaceae may, at times, be considered weedy. Many species occur in wet habitats and may clog waterways or occur in wet pastures. Most notable among the other genera in the family are *Carex* (Sedge), *Eleocharis* (Spikerush) and *Scirpus* (Bulrush).

Carex is a large genus of over 1,000 species represented in our area by nearly 150 species. It differs from other members of the family by having a sac-like structure completely enclosing the fruit. The flowers are all unisexual.

Eleocharis has leaves reduced to basal sheaths and bears a thickened triangular structure atop the fruit.

Scirpus is a variable genus. Unlike *Carex*, the flowers are not unisexual and the fruit is often surrounded by several bristles. The species illustrated here is of a round-stemmed, nearly leafless perennial, while other members of the genus may be annual, and some have 3-angled leafy stems.

a. *Carex* x 4/9
b. *Eleocharis* x 2/3
c. fruit x 10
d. *Scirpus* x 1

Rush Family: JUNCACEAE

Stems cylindrical or flattened; leaves 2- or 3-ranked, or sometimes reduced to basal sheaths only; flowers very small, consisting of 6 green or brown more or less papery segments; fruit splitting along three sides to release the seeds.

SOFT or COMMON RUSH

Juncus effusus L.

Tufted perennial from stout rootstocks; stems cylindrical; leaves reduced to chestnut-brown basal sheaths; flowers numerous, the flower cluster appearing to arise from 1 side of the stem and the single bract appearing as a continuation of the stem; fruit 3-angled with numerous minute seeds.

This is an extremely variable species with several varieties in our area; these occur in wet ground, mainly west of the Cascades.

TOAD RUSH

Juncus bufonius L.

Annual, 1 to 8 inches tall, rarely taller; leaf blades flat, less than 1/16 of an inch wide; stems branched; flowers borne singly or 2 or 3 together; fruit not angled, opening along 3 lines to release numerous minute seeds.

Wet ground throughout most of North America; in our area it is especially common as a lawn and garden weed west of the Cascades.

a. **SOFT RUSH**
Juncus effusus x 4/9

b. **TOAD RUSH**
Juncus bufonius x 2/3

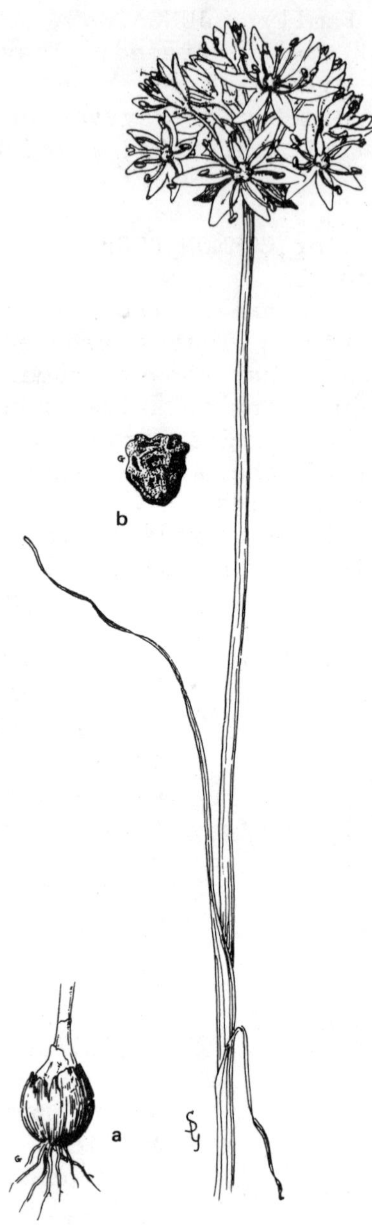

WILD ONION
Allium amplectens x 1 1/3
a. bulb x 1 1/3
b. seed x 4

Lily Family: LILIACEAE

This is a plant family containing many ornamental garden plants, a few vegetables including onion and asparagus, and an occasional weed.

Members of this family typically have showy flowers (some exceptions) with their parts in 3's or 6's (rarely 4's). The seeds are borne in dry or fleshy fruits in 3 compartments or attached along 3 lines in a single compartment.

WILD ONION

Allium amplectens Torr.

Perennial; bulb only one to a plant, with no offsets; bulb scales often purplish; leaves several, slender, but generally withered before flowering time; stem 3/4 to 1 1/2 feet tall, bearing an apical cluster of many small pinkish-white flowers, the flowers becoming papery as the seeds mature; seed black, 1/8 inch or larger, sharply angled.

West of the Cascades, and also to some extent east of them, this is a weed of grain fields, meadows, and pastures. Its perennial habit and abundant production of seeds make it difficult to control.

As in the case of Wild Garlic, though to a lesser degree, it taints dairy products by its strong onion flavor and therefore is detrimental in pastures and in hay.

Other native species of onion occur in the Pacific Northwest, but generally are confined to such restricted conditions of environment that they offer no problem to rancher or gardener.

WESTERN FALSE HELLEBORE

Veratrum californicum Dur.

Robust perennials from thick rootstocks; stems 3 to 7 feet tall; leaves 6 to 12 inches long, sheathing at the base; flowers in a branching panicle; flowers white to yellowish or even greenish; fruit 3/4 to 1 1/4 inches long with yellowish winged seeds.

Wet to moist areas, even at high elevations in the mountains and on both sides of the Cascades. This species is reportedly toxic to livestock and to humans. It is the cause of "monkey-face," a malformation of lambs.

WHITE or GREEN FALSE HELLEBORE

Veratrum viride Ait.

Similar to *Veratrum californicum*, but with yellow-green to deep green flowers in more or less drooping panicles.

Wet areas at medium or high altitudes.

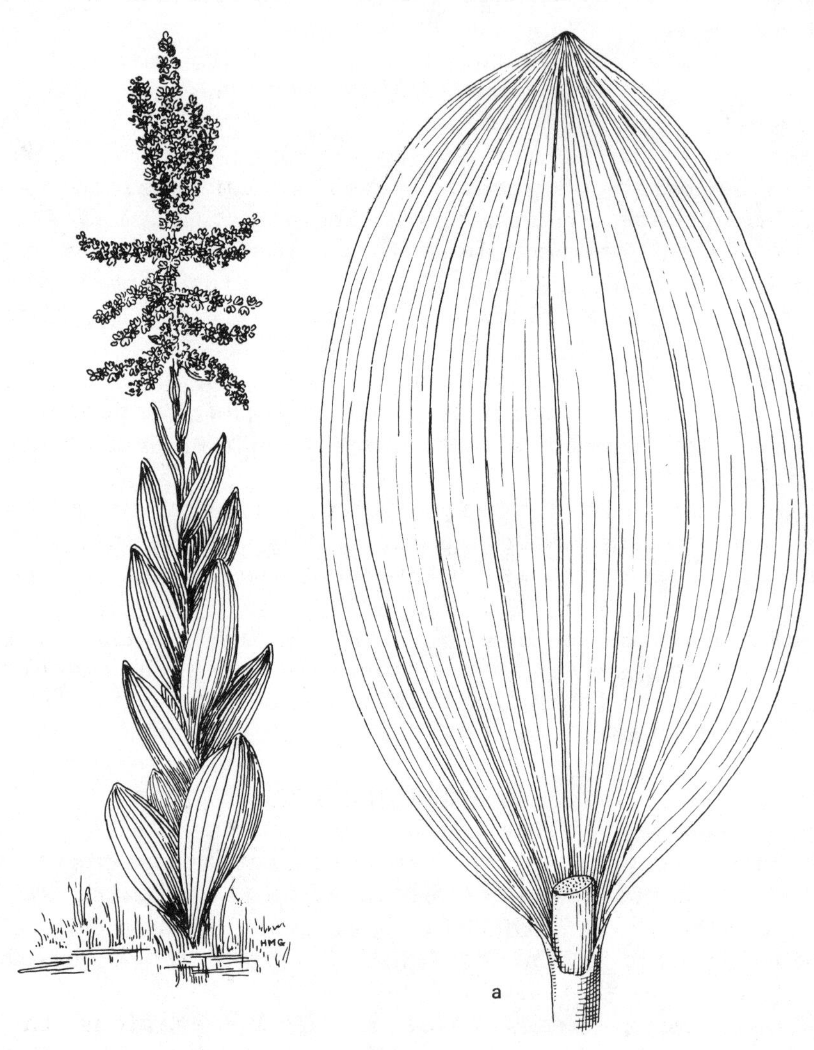

WESTERN FALSE HELLEBORE
Veratrum californicum x 1/10
a. leaf x 1/2

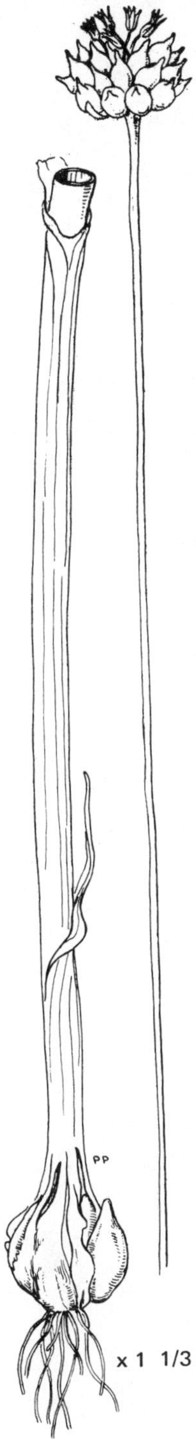

WILD GARLIC

Allium vineale L.

Perennial; offsets from the main bulb often forming masses at base of the plant, the covering of the bulbs papery and brittle; leaves several from near the base of the stem, long-pointed; stem cylindrical, tall, bearing a terminal cluster of flowers (some, or all of these often modified into bulbils); flower cluster, when present, emerging from the split base of a papery bract.

This is one of the most pernicious introduced weeds in the Northwest. Its ability to propagate rapidly by three methods—bulb offsets beneath the ground, bulbils in the flower cluster, and seeds, enables it to cover an area within a short time and control is difficult. It is a European weed, but readily transported in contaminated seed and capable of becoming firmly established wherever conditions are favorable. It is particularly troublesome in cultivated fields, lawns and gardens.

When these plants are eaten by cows a strong disagreeable flavor is evident in their dairy products, and even in their meat. In grain fields the bulbils, which are approximately the size of wheat, may be harvested with the grain and taint the flour.

The offset bulbs from the base of the plant, and the frequent substitution of bulbils for flowers, readily distinguish this species from common Wild Onion (*Allium amplectens*). In some cases what normally would be the flower head is composed completely of a cluster of bulbils. These eventually drop and form new plants, or they sometimes germinate in position to give the head the appearance of a bushy mass of green grass seedlings.

PP

x 1 1/3

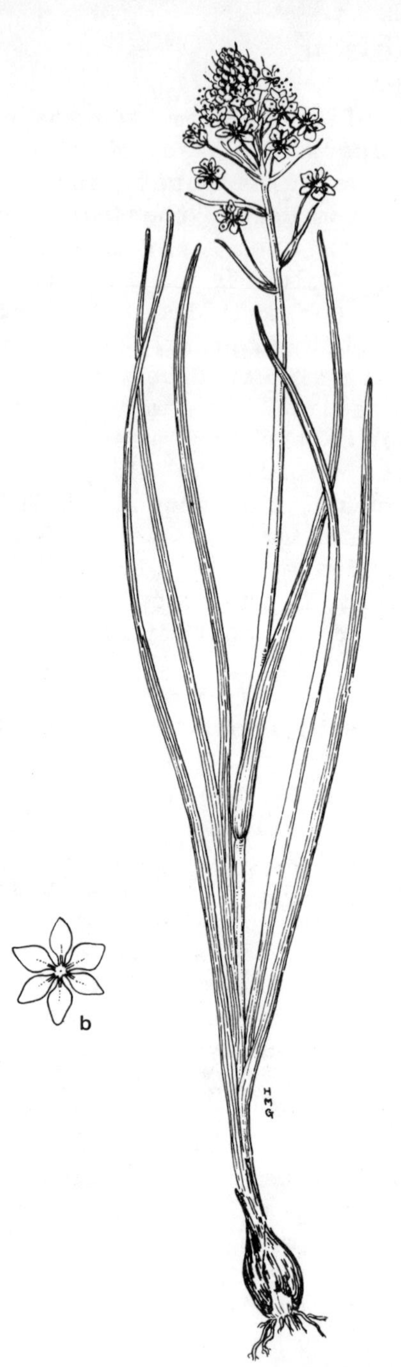

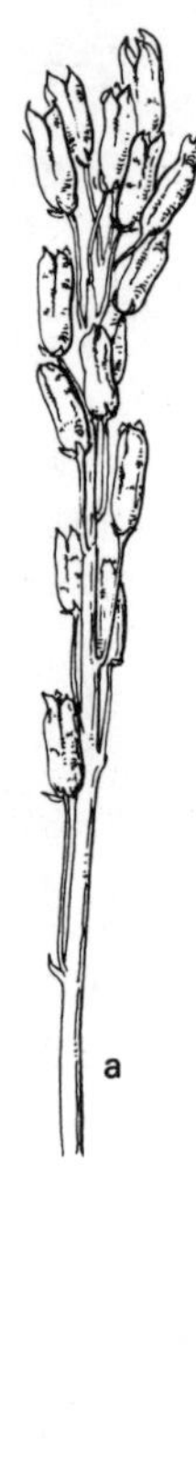

MEADOW DEATHCAMAS
Zigadenus venenosus x 1/2
a. fruits x 2/3
b. single flower x 1

MEADOW DEATHCAMAS

Zigadenus venenosus Wats.

Perennial by a dark-coated bulb; stem 3/4 to 2 feet or more tall; leaves several, narrow, nearly as long as the stem, generally rough-margined; flowers yellowish-white, in a cluster, at first dense and short, but elongating and becoming more open as the lower flowers mature; each flower spreading, 1/2 inch in diamter, 6-parted, each segment narrowed to a thick green gland at the base; fruit about 1/2 inch long, 3-lobed, narrow, opening by 3 valves, each valve bristle-pointed at the apex; seeds long, brown, several in each fruit.

This species of Deathcamas has a wide distribution. It is often found in low, wet pastures, but likewise occurs on rocky hillsides and dry plains. It is poisonous to stock and man, its toxic substance being distributed through all parts of the plant. Although this species is considered the most toxic of those found in our area, several others are sufficiently potent to be hazards on the range or in pastures. These include *Zigadenus elegans* Pursh (MOUNTAIN DEATHCAMAS), a large-flowered montane species, and *Zigadenus paniculatus* (Nutt.) Wats. (FOOTHILL DEATHCAMAS), common in pine forests and sagebrush deserts east of the Cascades.

These plants are not to be confused with Meadow Camas (*Camassia*) the species of which are most commonly blue-flowered and are harmless. Even the occasional white-flowered Meadow Camas occurring in blue-flowered colonies, or the white and pale blue color forms which occur abundantly in Douglas County, Oregon, and southward, can readily be distinguished from Deathcamas. The flowers of both are 6-parted; but that of Deathcamas spreads somewhat wheel-shaped, with a diameter of approximately 1/2 inch, and each segment is abruptly narrowed below the middle to form a thick green stalk-like gland. The flower of Meadow Camas, on the other hand, is somewhat bell-shaped at first and would flatten to a diameter of 1 1/2 to 2 or more inches. No glands occur at the base. An additional distinguishing character is found in the leaves of Deathcamas, which are deeply furrowed, giving them a distinct V-shape in cross section, while those of Meadow Camas are nearly flat.

Iris Family: IRIDACEAE

Perennial herbs with 2-ranked, sheathing leaves and showy flowers; ovary inferior.

ROCKY MOUNTAIN IRIS

Iris missouriensis Nutt.

Perennial from stout, branching rootstock; stems 1 to 2 feet tall, nearly leafless; basal leaves 1/4 to 1/2 inch wide; flowers 1 to 4 to a stem, blue-violet, 2 to 3 inches long.

Widespead east of the Cascades in wet meadows, along streambanks and in seepage areas in rangeland; west of the Cascades in the Puget Sound area of Washington. This species is reportedly toxic to cattle.

Nettle Family: URTICACEAE

This family is represented in our area principally by the nettles (*Urtica*). These plants are characterized by leaves paired on the stem, by small racemes of greenish flowers clustered in the leaf axils, and by stinging hairs. Another member of the family (*Parietaria pensylvanica* Muhl., PENNSYLVANIA PELLITORY) occurs east of the Cascades, but is unimportant as a weed. Its leaves are borne singly and are without stinging hairs.

STINGING NETTLE
Urtica dioica x 1/3

STINGING NETTLE

Urtica dioica L.

Perennial; stem angled, erect, bristly, 2 to 9 feet tall; leaves paired on the stem, broadly to narrowly ovate, coarsely toothed, smooth or hairy, stalked; flowers of 2 kinds, staminate and pistillate generally borne on different plants, both kinds small, greenish, borne on slender branches clustered in the upper leaf axils.

A second species, *Urtica urens* L. (BURNING NETTLE or DOG NETTLE), is occasional in the Pacific Northwest; it is an annual and is less than 2 feet in height.

Because of its stinging hairs, Nettle is one of our best-known native plants. It inhabits principally moist places along streams, preferably in areas somewhat shaded, where it often forms dense impenetrable stands. In gardens, to which it may be transferred in river-bottom loam, it is an unpleasant addition to the weed population.

RED SORREL
Rumex acetosella
a. inflorescence from pistillate plant x 1/3
b. staminate plant x 1/3
c. pistillate flower x 6
d. fruit x 6

Buckwheat Family: POLYGONACEAE

This family, to which the docks, smartweeds, and knotweeds belong, contains also Buckwheat and Rhubarb as important economic species. In most members of the family the joints of the stem are swollen and conspicuous with, in many cases, papery sheaths projecting above the leaf bases.

RED SORREL or SOUR DOCK

Rumex acetosella L.

Perennial, with slender creeping rootstocks; stem somewhat woody at the base, 1/2 to 2 feet tall, generally little branched; blade of the lower leaves somewhat arrowshaped with one or two conspicuous basal lobes, leaf stalk slender and bearing silvery scales at point of attachment with the stem, upper leaves often more slender and without lobes; leaves and stems very acrid to the taste; flowers borne in large branched clusters, the staminate and pistillate on different plants; staminate cluster larger, orange-yellow, bearing quantities of yellow pollen released in clouds under favorable conditions; pistillate clusters red-orange; fruit small, 3-angled, in color like polished mahogany, enclosed in 3 reddish persistent flower parts.

In the Pacific Northwest this European weed, long ago introduced into America, is abundant on burns and elsewhere. While apparently thriving best on acid soil, it adapts itself to other types of conditions, often occurring as a weed in lawns, fields, gardens, and along roadsides.

The flower of all true docks has a 6-parted perianth (sepals in this case). The outer 3 segments are small and remain unchanged, but the inner 3 enlarge as the fruit matures, and in many species bear swollen, cork-like or grain-like bodies (tubercles) on the outer surface of the midvein. The presence or absence of these bodies on some or all of the segments has become important in identifying species. The three inner segments also may become conspicuously wing-like and veiny in certain species, and some are characterized by marginal teeth or bristles.

CURLY DOCK
Rumex crispus
a. lower leaf x 2/3
b. calyx enclosing mature fruit x 4
c. fruit x 8

CURLY DOCK

Rumex crispus L.

Perennial; stems 2 to 5 feet tall, stout, slightly ridged, often reddish, generally unbranched below; leaves elongated, crisped on the margins, the lower long-stalked, the upper leaves shorter; flowers greenish, numerous, crowded in a long thick branching cluster, the upper half or third unbranched, at maturity becoming dense, reddish-brown, the long compact spike-like cluster standing conspicuously, even through the winter, gradually weathering to grayish brown; individual fruits enclosed in veiny wing-like structures 1/8 to 3/16 inch long, each wing or sometimes only one or two bearing a thickened corky tubercle; fruit stalked, the stalks jointed.

A native of Europe which has become widely distributed and abundant, this is one of our most common weeds. The rust-colored plants are conspicuous in the winter landscape on both sides of the Cascades, many of the fruits remaining attached for several months, but gradually loosening their hold and becoming distributed by the winter winds.

BROADLEAF DOCK

Rumex obtusifolius L.

Perennial; stem slender, erect, 1 1/2 to 4 feet tall; leaf blade 4 to 9 inches long, oval, ovate, or lance-shaped, the base rounded or heart-shaped, the apex rounded or somewhat pointed, the leaf stalk long; flowering panicles large; flowers at maturity elongated, the 3 wings (inner perianth parts) strongly veiny, only one generally bearing a tubercle, but each with 1 to 3 pointed teeth on each side; fruit 3-angled, shining, pale brown.

Broadleaf Dock is a native of Eurasia but has become a common weed in pastures and fields west of the Cascades. This and other docks are troublesome in crops of clover grown for seed, since the docks produce quantities of seeds which are harvested with those of clover.

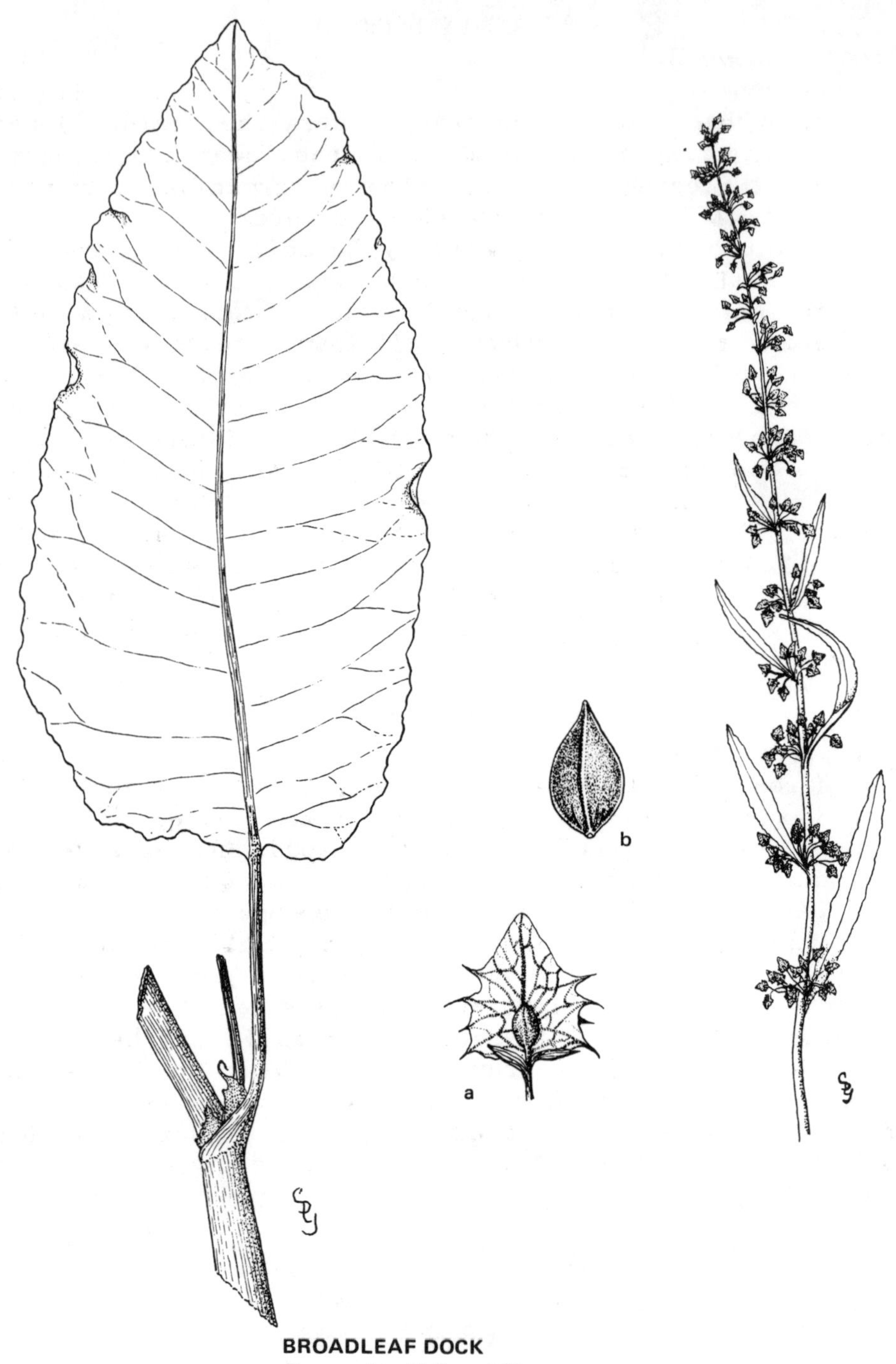

BROADLEAF DOCK
Rumex obtusifolius x 2/3
a. calyx enclosing mature fruit x 4
b. fruit x 8

WILD BUCKWHEAT or BLACK BINDWEED

Polygonum convolvulus L.

Annual; stems weak, twining or trailing, 3/4 to 3 feet long; leaves broadly arrow-shaped, the blades 2/3 to 2 inches or more in length, with slender stalks arising from the stem together with a short papery sheath; flowers greenish, clustered in the leaf axils; fruit 3-angled.

A native of Europe, it is now known in practically the entire United States as a weed of gardens, orchards, and waste places.

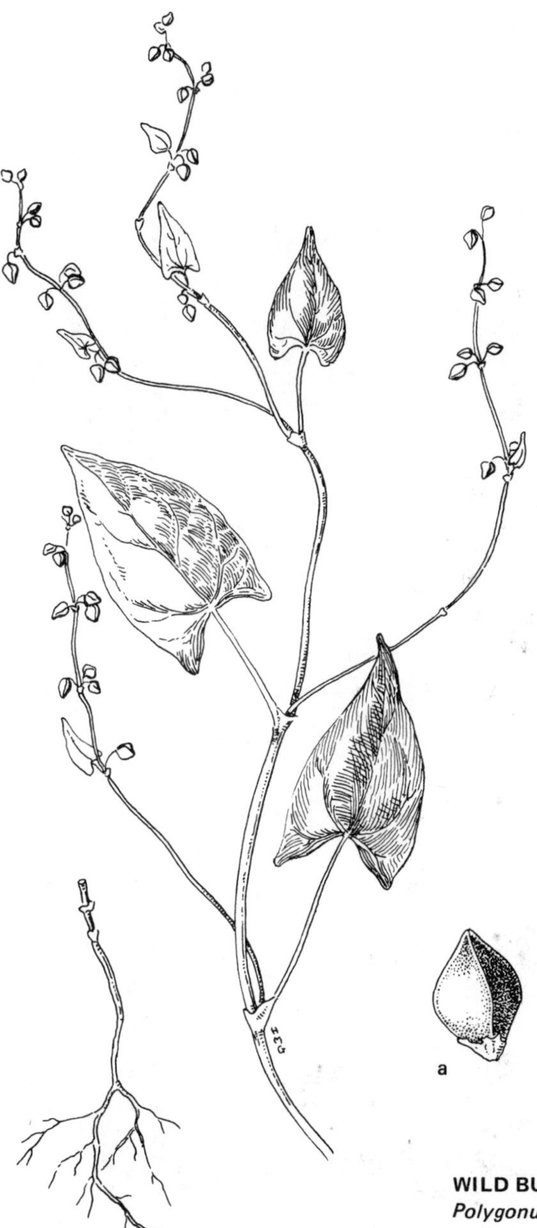

a

WILD BUCKWHEAT
Polygonum convolvulus x 2/3
a. fruit x 6

LADYSTHUMB

d. upper portion of plant x 2/3

e-f. variations of fruit x 10

e′-f′. cross-section of e and f x 10

PROSTRATE KNOTWEED

Polygonum aviculare

a. plant x 2/3

b. axillary cluster of flower buds x 1

c. fruit x 10

PROSTRATE KNOTWEED

Polygonum aviculare L.

Extremely variable annual; stems wiry, much branched, erect if competition for sunlight is keen, otherwise (and usually) forming wide-spreading mats; leaf blade 1/2 to 1 1/2 inches long, generally narrow, short-stalked, each leaf accompanied by a torn silvery sheath at its base; flowers very small, greenish or pinkish, clustered in the leaf axils; fruit dark brown, 3-angled, shining.

Prostrate Knotweed, which is said to accompany civilized man wherever he goes, is a native of the Old World but is now found abundantly in America and elsewhere. It is somewhat troublesome in gardens and strawberry fields but more typically is encountered in hard-packed paths, roadsides, waste places, and farmyards.

LADYSTHUMB or SMARTWEED

Polygonum persicaria L.

Annual; stems usually stout with knobby joints, erect or more often spreading, 1 to 3 or more feet long; leaves rather narrow, smooth or sparsely hairy, the margins bristly; flowers bright or pale pink, borne in short thick spikes at stem tips and in the leaf axils; fruit 3-angled.

This European species has long been a resident of low moist land in the Northwest. In recent years it has become a troublesome weed of cultivated fields and irrigated pastures. This species is found throughout North America, but is more abundant west of the Cascades in our area.

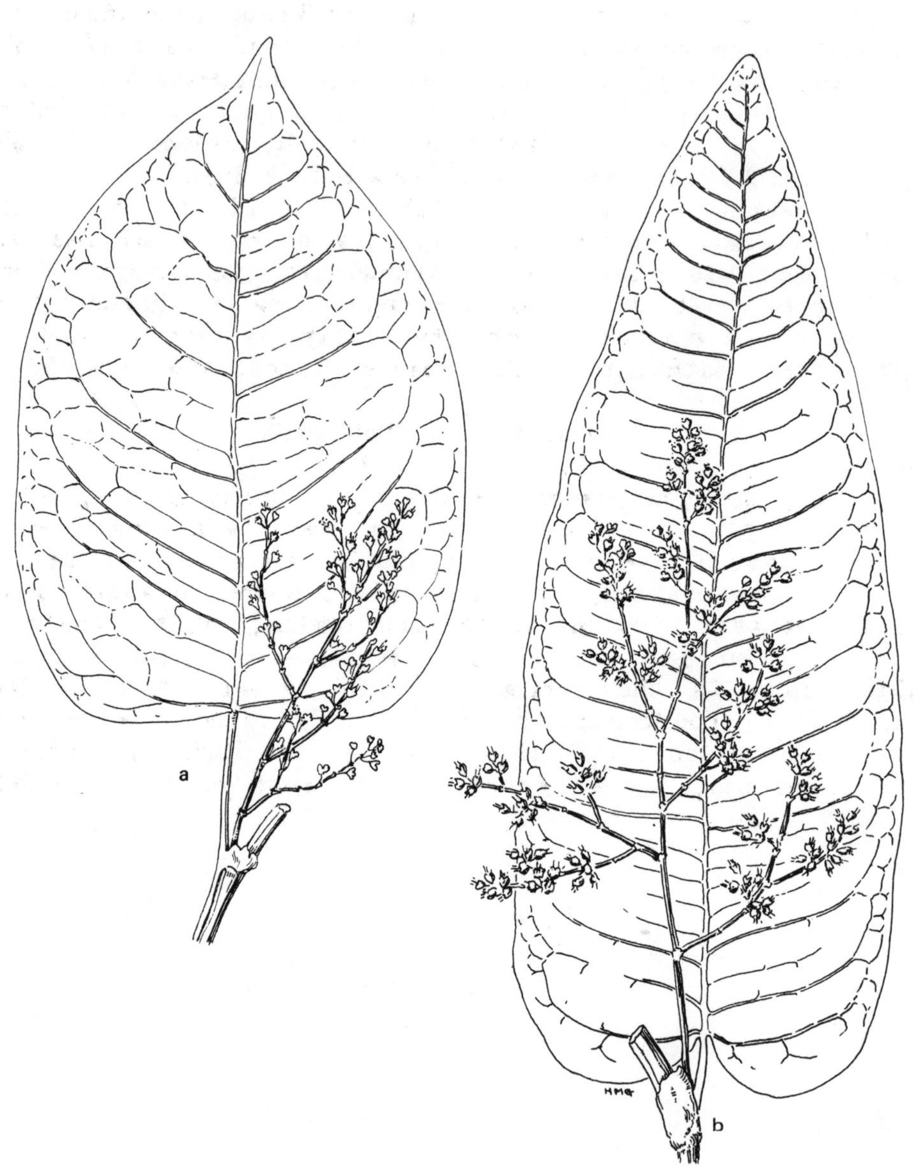

a. **JAPANESE KNOTWEED**
Polygonum cuspidatum x 2/3

b. **SAKHALIN KNOTWEED**
Polygonum sachalinense x 2/3

JAPANESE KNOTWEED or FLEECEFLOWER

Polygonum cuspidatum Sieb. & Zucc.

Perennial; stem stout, 4 to 9 feet tall, woody but dying at end of growing season; leaves short-stalked, broadly ovate, 2 to 6 inches long and about 2/3's as broad, base slightly heart-shaped, wedge-shaped or straight, apex abruptly narrowed to a point; papery sheaths below the leaf bases soon disappearing; flowers creamy white, borne in large plume-like clusters at ends of stems and in leaf axils.

An Asiatic species introduced as a garden ornamental, this plant has escaped to roadsides and cultivated fields of several counties in the far West, and is occasionally reported as a pest.

SAKHALIN KNOTWEED or SACALINE

Polygonum sachalinense Schmidt ex Maxim

Similar to *Polygonum cuspidatum* but often reaching 12 feet in height; the leaves narrower and generally longer (4 to 12 inches in length), more or less heart-shaped at the base; the papery sheaths usually remain, brown and ragged, on the stem.

Sakhalin Knotweed was introduced experimentally into this country some years ago as a forage plant as well as an ornamental. As the latter, it has merit, but it seems to have failed as a practical source of forage. In addition, its tendency to spread by stout rootstocks, and its tenacity when once established, make it a potential nuisance. It is frequently so reported in scattered sections of the Pacific States, mainly west of the Cascades.

HALOGETON

Halogeton glomeratus x 1/2

a. leaf x 3

b. portion of inflorescence x 4/5

c. single flower x 5

Goosefoot Family: CHENOPODIACEAE

To this family, which contains spinach and beet, belong also a number of wayside weeds. Some of these are common only in alkaline soils while a few become troublesome in gardens, cultivated fields and grazing lands.

The flowers of all our species are small, inconspicuous, generally greenish, and borne in clusters at the ends of stems or in the leaf axils. The seeds are often black and shining. The leaves and stems of many members are typically mealy.

HALOGETON

Halogeton glomeratus C. A. Mey.

Annual; branching from the base, the stems spreading at first then becoming erect, from a few inches to a foot or more long, generally with numerous short secondary branches; leaves thick, nearly tubular, abruptly ending in a slender needle-like point; flowers greenish, inconspicuous, of two kinds--the larger with wing-tipped sepals surrounding the fruit, the other with sepals tooth-like at the apex; seed more or less flattened, the spiral form of the embryo clearly evident.

In early stages it somewhat resembles Russian Thistle and may therefore pass unnoticed. It is readily distinguished by its leaves, which remain unchanged during the season's growth, whereas Russian Thistle bears two distinct successive forms. Also, in Halogeton, tufts of white cottony hairs occur in the leaf axils, while these are absent in Russian Thistle. When the seeds are ripe, the side-winged calyces enclosing fruits, crowded densely on the stems, give the plant a thick fluffy appearance in strong contrast to the slender spiny seed-bearing plants of Russian Thistle.

Both these species may become tumbleweeds, both may turn yellow or reddish in late summer, and both normally produce an abundance of seed. Both are troublesome weeds, reproducing heavily and crowding out valuable range grasses or the crops of cultivated fields. But Halogeton's reputation as a sheep destroyer makes it a far more formidable enemy on the range.

FIVEHOOK BASSIA
Bassia hyssopifolia x 1
a. Calyx enclosing fruit x 10

FIVEHOOK or HYSSOP BASSIA

Bassia hyssopifolia (Pall.) Ktze.

Coarse annual; 2 to 6 feet tall; stem erect, much branched, with a covering, at least above, of fine white hairs; leaves entire, narrow, silky-hairy; flowering spikes slender, dense, white-woolly; fruit enclosed in a 5-lobed calyx, each lobe bearing a small stiff hooked spine.

This Eurasian weed was introduced into America in comparatively recent years, but is now abundant in irrigated regions of eastern Oregon, Washington, and in Idaho. In some areas it has become a pernicious pest while in others it is reported as palatable to stock, furnishing feed of considerable value, particularly after maturity of the fruit.

SLIMLEAF LAMBSQUARTERS

Chenopodium leptophyllum (Moq.) Wats.

Annual; stems 8 inches to 2 1/2 feet tall; whole plant usually gray and mealy, but occasionally somewhat greenish; leaves 3/8 to 2 inches long, less than 1/8 of an inch wide, 1- to 3-veined; flowers minute in numerous short spikes.

Desert areas east of the Cascades.

MEXICAN TEA

Chenopodium ambrosioides L.

Annual or short-lived perennial; stems usually much branched, more or less "hairy' and covered with unstalked yellow glands; leaves irregularly toothed, the upper leaves smaller and less toothed to nearly entire; flowers minute, borne in numerous short spikes.

Moist ground, especially common in gravelly flood plains; thought to be introduced from Mexico or South America.

COMMON LAMBSQUARTERS

Chenopodium album L.

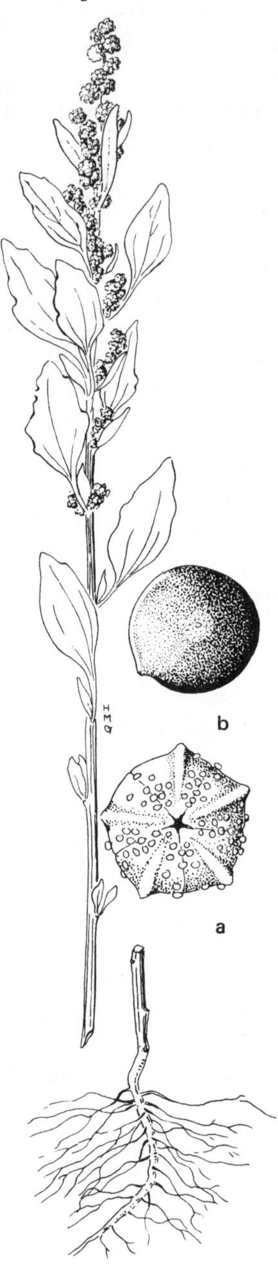

Extremely variable annual; stems ridged, erect, 1 to 4 or more feet tall, somewhat branching, leaves and stems more or less grayish mealy; leaf blades 1/2 to several inches long, broadly or narrowly ovate, the lower often wavy-margined or somewhat toothed or lobed, the upper generally narrower and often entire, the leaf stalk approximately 1/2 as long as the blade; flowers very small and inconspicuous, greenish, mealy, borne in small clusters arranged in a long spike or panicle; seeds numerous, black and shining, flattened horizontally, the fruit loosely enclosed in the 5-parted, 5-crested calyx.

Naturalized from Eurasia, this is a common weed of cultivated fields, gardens, waste places and about old buildings. An effective way to combat it is by table use, either as greens or salad. In the young stages it is tender and succulent and often ready for use before garden vegetables have made sufficient growth for this purpose. Belonging to the same family as beets and spinach, it provides comparable qualities in food value and flavor.

In addition to being a garden pest, it is a host of the beet leafhopper and therefore undesirable in the vicinity of beet crops.

COMMON LAMBSQUARTERS
Chenopodium album x 1/2
a. calyx enclosing mature fruit x 15
b. seed x 15

RED ORACH

Atriplex rosea L.

Annual, usually much branched; 8 inches to 4 feet tall, rarely taller; gray-mealy at least on the under surfaces of the leaves; leaves 3/4 to 2 inches long, coarsely-toothed, or the upper entire; flowers small, in short spikes.

Introduced from Eurasia; now widespread east of the Cascades, often in alkaline areas.

SPREADING ORACH

Atriplex patula L.

Much branched annual to 4 1/2 feet tall; leaves usually somewhat mealy, the lower leaves opposite, the upper alternate; leaf shape extremely variable; flowers small; clustered in numerous spikes.

Several varieties occur in our area; usually in saline or alkaline soil along the coast and east of the Cascades.

GARDEN ORACH

Atriplex hortensis L.

Annual; 2 to 8 feet tall, much branched; leaf blades more or less triangular in outline, 2 to 8 inches long and half as broad; flowers small, borne in numerous short spikes; fruit broadly-winged, bearing a single seed.

This species is native of Asia and is occasionally cultivated and escaped and becoming weedy in our area.

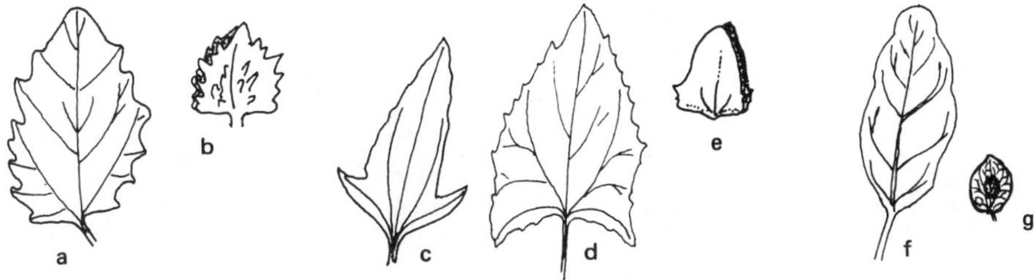

RED ORACH
Atriplex rosea
a. leaf x 2/3
b. bracts enclosing the fruit x 3

GARDEN ORACH
Atriplex hortensis
f. leaf x 1/4
g. bracts enclosing the fruit x 1/2

SPREADING ORACH
Atriplex patula
c-d. variation in leaf shape x 2/3
e. bracts enclosing the fruit x 3

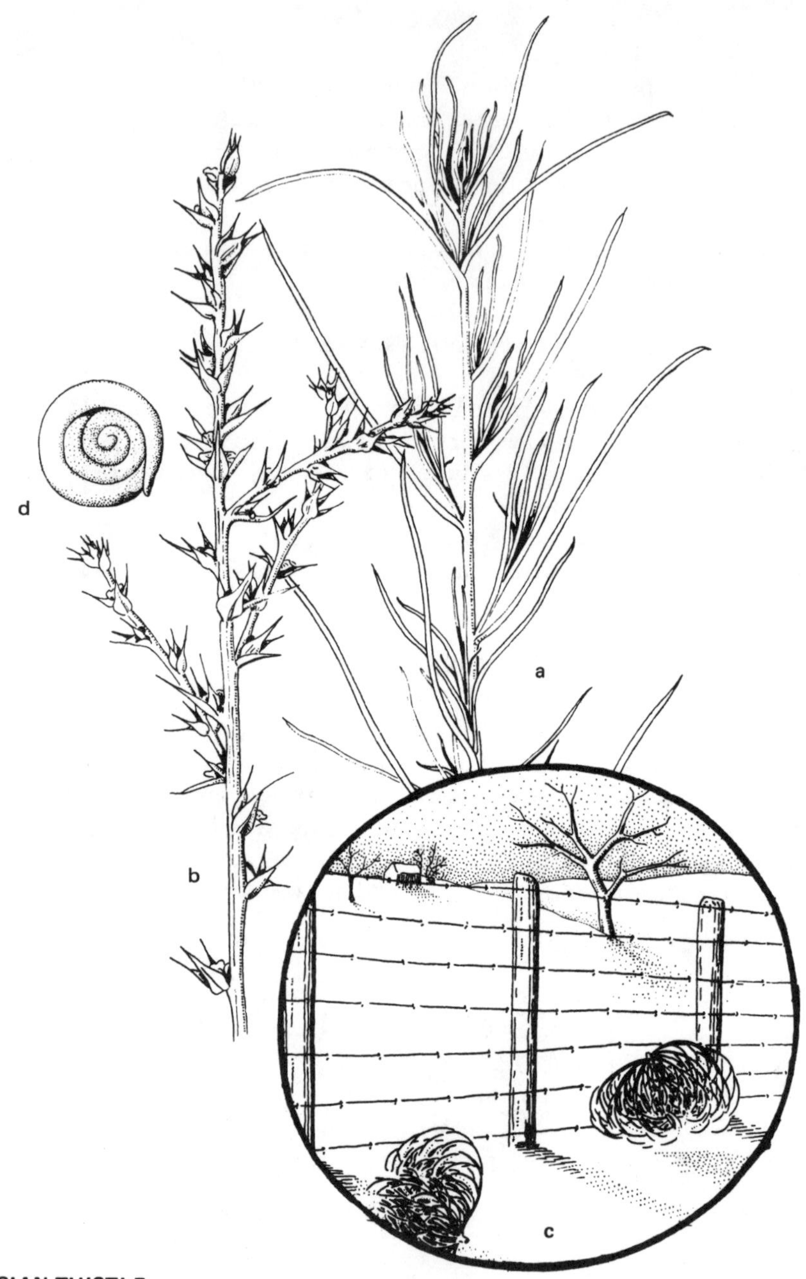

RUSSIAN THISTLE
Salsola kali var. *tenuifolia*
a. spring condition x 1 1/3
b. summer condition 1 1/3
c. fall and winter condition
d. seed x 6

RUSSIAN THISTLE

Salsola kali L. var. *tenuifolia* Tausch

Annual; in early stages or on poor soil plant more or less prostrate with ends of the branches turned upward; at maturity or on better soil, the plants 1/2 to 2 feet tall, bushy with spreading branches; stems ridged, pale green or often reddish; first leaves long, soft, thread-like; later leaves much shorter, stiff, scale-like, spine-tipped; flowers small, greenish, borne in axils of the upper leaves; fruit enclosed in the persistent calyx, this becoming papery at maturity, with a horizontal circular wing; seed distribution effected by the entire plant breaking off at the ground line, drying with incurved branches, thus assuming a ball-like form readily driven before the wind and sowing its seed as it rolls along.

This weed is said to have been introduced into the Middle West in 1874 in flax seed imported from Russia. From here it spread westward, and has long been a pest on the plains east of the Cascades. Although an annual and therefore presumably offering little difficulty in control, each plant is capable of producing thousands of seeds. This, together with its effective means of superintending its own distribution (the tumbleweed habit), has made of it a formidable enemy.

When in the very young succulent stage, the plant is edible and may be used for greens or salad.

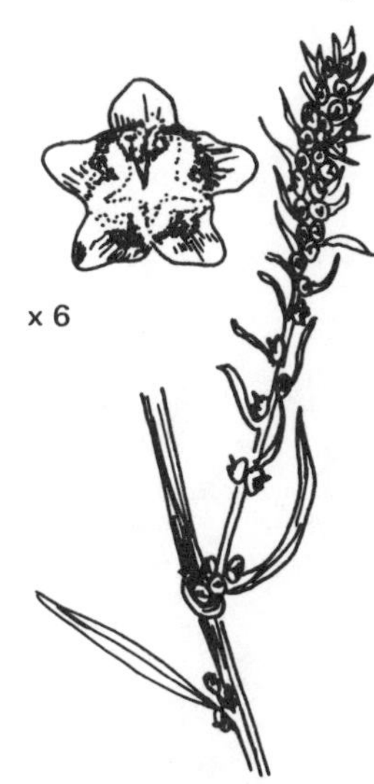

KOCHIA

Kochia scoparia (L.) Schrad.

Annual; stems much branched, usually "soft-hairy" but occasionally smooth, reddish-tinged, at least in the Fall; leaves 1/2 to 2 1/4 inches long, 1/16 to 1/4 of an inch wide, the margins ciliate, the upper surface smooth, the under surface usually with soft rust-colored hairs; leaf blades with 3 or 5 prominent veins; flowers in small axillary clusters.

Introduced from Eurasia, becoming weedy in irrigated areas east of the Cascades. Some horticultural forms of this species are sold under the names of Summer or Mock Cypress and Red Belvedere.

REDROOT PIGWEED
Amaranthus retroflexus x 1/2
a. single flower x 6
b. young fruit x 8
c. seed x 8

Redroot Pigweed Family: AMARANTHACEAE

This family is represented in our area by several species, all of which are widely distributed weeds. Aside from its weedy members, the family is little known horticulturally except for the garden ornamental plant called Cockscomb.

In all our species the flowers are minute and borne in small or large often plume-like clusters. Intermixed with the flowers are small stiff bracts which give a harsh feel to the cluster. Generally the flower clusters are greenish, but in Cockscomb the plumes may be red, yellow, white, or pink.

REDROOT PIGWEED

Amaranthus retroflexus L.

Annual or winter annual; stem erect, 1 inch to 4 feet tall, woody in larger plants, rough-hairy to smooth, ridged, often reddish especially below the ground line, the red color continuing down the tap root; leaves long-stalked, the blades more or less broadly ovate, veiny, with somewhat wavy margins; flower clusters generally branched, dense, borne both at apex of the stem and in axils of leaves, coarse and rough from the presence of numerous small stiff and pointed scales; plants bearing either staminate or pistillate flowers; seeds abundant, rounded, black, shining, each at first enclosed in a papery sac, the apex of this sac later splitting off in the form of a cap, thus allowing the seed to escape. (A hand lens is needed to see this character clearly.)

Introduced from tropical America, this species has now become widely established in the United States. It is particularly troublesome in cultivated ground such as gardens. The ability of its seeds to mature early and germinate immediately insures a succession of crops during a single growing season and necessitates constant cultivation. Although the plants, being annuals, are easily removed, the seeds remaining in the ground are a constant menace. This situation is aggravated by the fact that seedlings produced in rapid succession during the summer may bear flowers and fertile seed when only an inch or two tall.

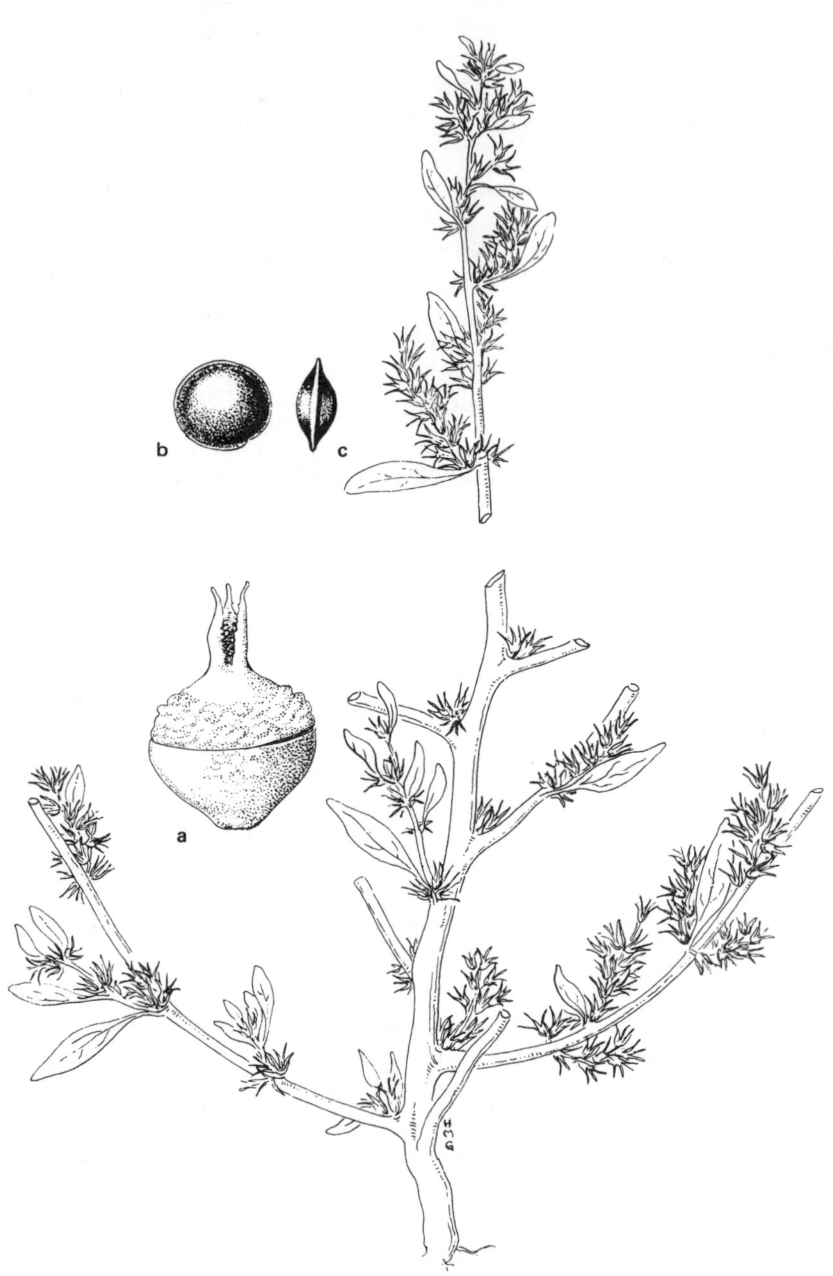

TUMBLE PIGWEED
Amaranthus graecizans x 2/3
a. fruit x 14
b-c. seeds x 12

TUMBLE PIGWEED

Amaranthus graecizans L.

Annual; ascending or more or less prostrate, much branched, 1 to 3 feet tall or broad; leaf blade narrowly ovate or obovate, rounded or notched at the apex, pale green above, sometimes reddish beneath, narrowed at the base; flowering spikes short, with small spine-tipped scales; plants bearing both staminate and pistillate flowers; seeds very small, dark reddish brown, shining, enclosed in a strongly wrinkled covering; seed distributed by the tumbleweed habit.

Like Redroot Pigweed, this species has been introduced from tropical America, and its methods of seed dispersal is similar. It is most common east of the Cascades where it occurs in desert areas; west of the Cascades it occurs as a weed of cultivated or waste land.

In addition to *Amaranthus retroflexus* and *Amaranthus graecizans*, two other species may be encountered as weeds in our area. *Amaranthus albus* L. (Tumble Pigweed) is similar to *Amaranthus graecizans* in that its flowers are all borne in axillary clusters and differs from it in its erect habit. *Amaranthus powellii* Wats. (Powell Amaranth) is similar to *Amaranthus retroflexus* in that it has terminal spikes, as well as axillary clusters of flowers and differs from it in having both staminate and pistillate flowers on a single plant.

Carpetweed Family: AIZOACEAE

Our single weedy member of this family is the insignificant Carpetweed described below. Better known horticulturally is the showy genus *Mesembryanthemum*, which in warmer areas is used for a colorful ground cover and of which one native species occurs on the coast of southwestern Oregon.

CARPETWEED

Mollugo verticillata L.

Annual; branching, forming small mats on the ground; leaves whorled, narrow, broader at the apex, narrowed to a stalk-like base; flowers very small, slender-stalked, borne in the leaf axils; seeds abundant, minute, red-brown, more or less striped and ridged dorsally.

This is a naturalized tropical weed not uncommon in moist sandy areas throughout temperate North America. It is frequently introduced into gardens by way of river sand.

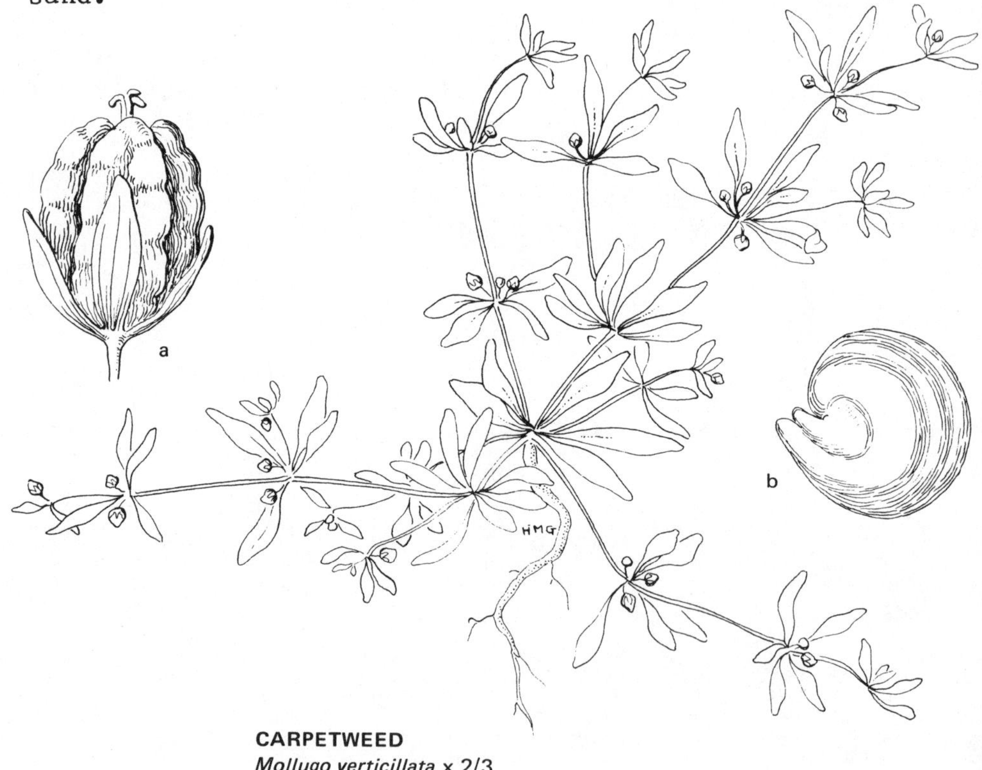

CARPETWEED
Mollugo verticillata x 2/3
a. fruit x 8
b. seed x 32

Purslane Family: PORTULACACEAE

This is a family common in the Pacific Northwest, but with only a few members reported as weeds. To the family belong several showy *Lewisia* species, and garden *Portulaca*.

MINER'S LETTUCE
Montia perfoliata x 2/3

MINER'S LETTUCE

Montia perfoliata (Donn) Howell
(=*Claytonia perfoliata* Donn)

Low fleshy annual; basal leaves varying from narrow and strap-shaped to long-stalked, broad-bladed forms; stem leaves a single pair united to form a rounded or 2-angled disk below the flowers; flowers white or pink; seeds usually 3, shiny black.

This species is extremely variable in height and in the size and shape of the leaves.

Usually in moist, shady areas on both sides of the Cascades.

PALE MINER'S LETTUCE

Montia spathulata (Dougl.) Howell
(=*Claytonia spathulata* Dougl. ex Hook.)

Fleshy annual, 3 inches or less in height; basal leaves strap-shaped, stem leaves 2, usually united on one side for less than half their length; flowers small, white to pink; seeds 2 or 3, black and shining.

Often in drier sites than *Montia perfoliata*.

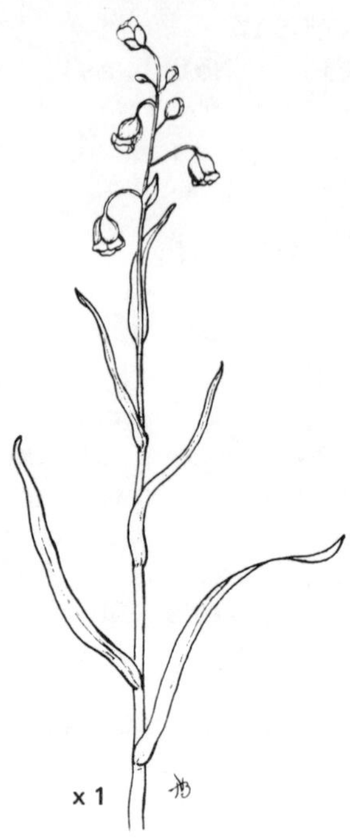

NARROWLEAF MINER'S LETTUCE

Montia linearis (Dougl.) Greene
(=*Claytonia linearis* Dougl.)

Annual; stems to 8 inches in height, often much branched; leaves alternate, narrowly linear; flowers in loose, often one-sided racemes; petals white, from shorter than the sepals to slightly exceeding them; fruit splitting at maturity to release the shiny black seeds.

This native species occurs on both sides of the Cascades from British Columbia through California and east to Montana and Utah. It is usually found in moist sandy soil and is sometimes weedy in cultivated fields.

REDMAIDS

Calandrinia ciliata (R. & P.) DC. var. *menziesii* (Hook.) Macbr.
(=*Calandrinia caulescens* of some authors, but not H.B.K.)

Somewhat fleshy annual; stems low and spreading or more or less erect; leaves narrow and strap-shaped; flowers borne in the axils of the upper leaves, petals rose-red or rarely white; seeds numerous, black and shiny.

Although this is a native species growing in areas that are moist, at least early in the spring, it often becomes weedy in cultivated fields and orchards.

COMMON PURSLANE

Portulaca oleracea L.

Annual; smooth, fleshy, branched from the base, prostrate or with branch tips ascending, stems often reddish; leaves paired, broadest near the apex, narrowed at the base; flowers in small clusters, yellow, opening only in the sunshine; sepals 2, fused below with the fruit; petals 5; fruit splitting horizontally at maturity, the upper part separating as a lid; seeds flattened, dull, minutely granular.

A native of Asia or Europe and troublesome in much of the United States, including California. It is reported from both east and west of the Cascades and should be treated as an undesirable invader. Its common name, Purslane, is sometimes shortened colloquially to "Pusley."

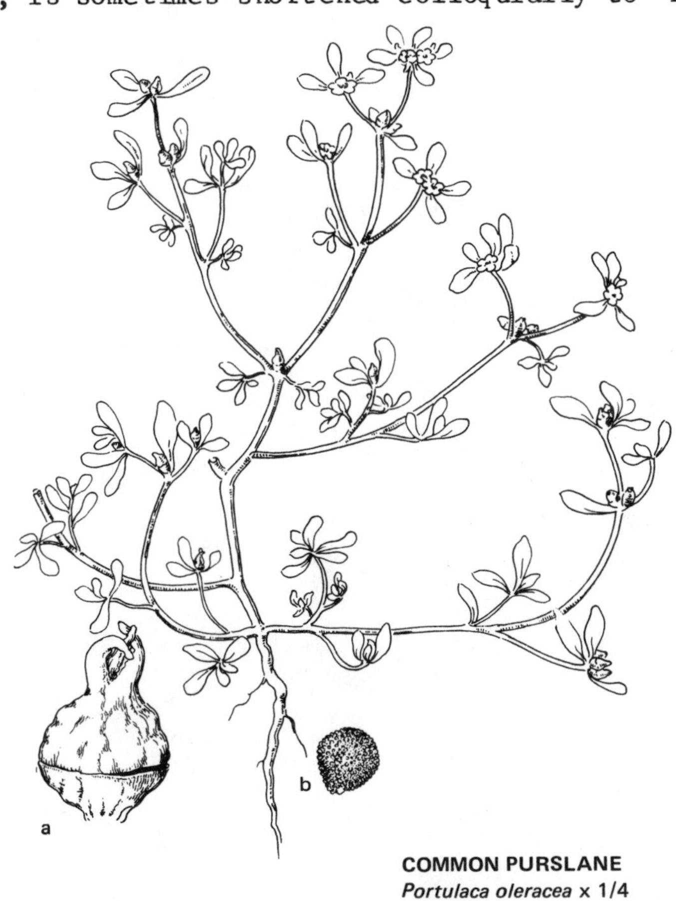

COMMON PURSLANE
Portulaca oleracea x 1/4
a. fruit x 4 1/2
b. seed x 10

BOUNCINGBET
Saponaria officinalis x 1

Pink Family: CARYOPHYLLACEAE

This family contains many ornamental garden species, notably Carnation, Pink, and Sweet William. To it also belong several troublesome weeds. Wide variation occurs in superficial characters but, in general, we find the outer flower parts in 4's or 5's, stamens of the same number or twice as many, and seeds borne in a dry fruit which splits at the apex to release the seeds.

BOUNCINGBET

Saponaria officinalis L.

Perennial with stout erect smooth branching stems, 1 1/2 to 3 feet tall, with swollen nodes; leaves smooth, 2 to 4 inches long, narrowly ovate or elliptic, pointed, short-stalked, with 3 distinct veins from the base; flowers conspicuous, generally pink, in crowded arrangement at ends of main stem and branches; calyx narrow; petals long-clawed at the base, slightly notched at the apex; fruit opening at the apex by 4 teeth; seeds many, somewhat curved, blackish, pebbled.

Introduced originally from Europe as a garden plant but now an escape from cultivation, Bouncingbet has become well established in a number of areas west of the Cascades. It is known as a stock-poisoning plant, but is so rarely grazed that its presence is of little significance. However, it spreads rapidly and replaces plants of greater value.

SLEEPY CATCHFLY

x 3

Silene antirrhina L.

Annual; stems 8 inches to 2 1/2 feet tall, with bands of sticky glands below some pairs of leaves; stem leaves narrow, 1 to 2 1/2 inches long, less than 1/2 inch wide; calyx 10-veined; flowers white or pink; fruit 3-chambered, surrounded by the calyx; fruit opening at the top by 8 to 10 teeth to release the numerous gray-black, pebbled seeds.

Weedy species in open, often sandy ground; abundant in temperate North America.

a. **WHITE COCKLE**
Lychnis alba x 1/2
b. seed x 12

NIGHTFLOWERING CATCHFLY
Silene noctiflora
c. upper portion of plant x 1/2
d. lower portion of plant x 1/2
e. seed x 12

WHITE COCKLE

Lychnis alba Mill.

Annual or more often biennial or a short-lived perennial; stem 1 1/2 to 4 feet tall, stiffly erect, branching, much enlarged at the nodes, minutely hairy; leaves short-hairy, basal leaves stalked, narrowed at both ends, stem leaves shorter, borne in pairs, the uppermost leaves rounded at the base, pointed at the apex; flowers white (rarely slightly pinkish), somewhat nodding, about 1 inch in diameter; calyx narrow at first, developing with the fruit into a bulbous structure 5-toothed at the apex; petals notched; fruit enclosed in the enlarged calyx, opening by teeth; seeds many, gray, minutely pebbled.

A European weed now widely distributed in parts of the Pacific Northwest; a pest in grain and other crops, or sometimes in relatively undisturbed sites. Even though stamens and pistils are generally borne on separate plants and only the pistillate plants are fertile, each flower of the latter is capable of producing such an abundance of seed, and each plant capable of producing so many flowers, that an infested field may become valueless.

NIGHTFLOWERING CATCHFLY

Silene noctiflora L.

Annual; stem generally unbranched, 1 to 3 feet tall; leaves rather broad, both stems and leaves generally covered by sticky hairs; flowers one or several at ends of stems, white, yellowish, or very pale pinkish, opening in the evening; seeds slightly curved, roughened by minute points.

Since this species reproduces only by seeds, it is less difficult to control than its relative, Bladder Campion, yet it is often troublesome in cultivated fields. Its seed is difficult to separate from that of the commercial clovers, especially alsike, and badly infested commercial seed is not marketable.

BLADDER CAMPION

Silene cucubalus Wibel

Perennial by branching woody rootstocks; stem branching, smooth; leaves smooth, ovate or lance-shaped, the margins generally not toothed; calyx at first slender, becoming greatly inflated, thin, veiny, and often purplish as the fruit matures, eventually becoming a veiny, papery sac-like structure about the bulbous fruit; fruit opening at the toothed apex to allow escape of numerous small grayish pebbled seeds.

This weed, introduced from Europe, is a serious pest in seed crops. Its ability to spread underground renders control difficult; and its seed, though differing widely in detailed appearance from that of alsike and certain other clovers, is of so nearly the same size that separation is practically impossible.

BLADDER CAMPION
Silene cucubalus x 1/2
a. seed x 8

COW COCKLE or COW HERB

Vaccaria segetalis (Neck.) Garcke ex Asch.
(=*Saponaria vaccaria* L.)

Gray-green annual; stems 1 1/2 to 3 feet tall, widely branching above, each branch ending in a flower; leaves narrow or broad, wider at the base, pointed at the apex, smooth; flower small but conspicuous, with inflated calyx and toothed pink petals; fruit at maturity remaining in the enlarged calyx; seeds gray, pebbled.

A European plant widely distributed east of the Cascades, and a conspicuous roadside and grainfield weed. The seeds, which are poisonous to stock, may be harvested with the grain in an infested field, resulting in serious contamination.

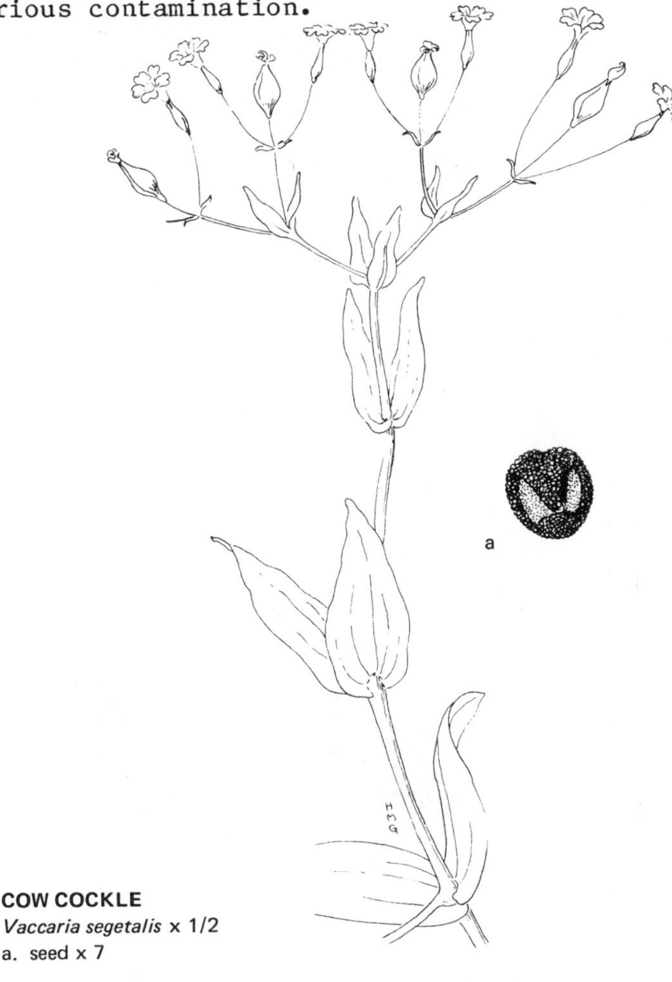

COW COCKLE
Vaccaria segetalis x 1/2
a. seed x 7

CORN SPURRY

Spergula arvensis L.

Annual; much branched at the base, branches 1/2 to 2 feet long, somewhat sticky, erect or more or less spreading; leaves narrow, cord-like, fleshy, arranged in a series of apparent whorls at the stem joints; flowers small, white, in loose clusters at ends of the branches; fruit small; seeds somewhat flattened, dull black, usually with minute whitish warts and a circular whitish wing.

Introduced from Europe, this plant has become common in parts of the Northwest, and is sometimes troublesome in gardens, orchards, and cultivated fields. It may be found in sandy waste places both east and west of the Cascades, but is more common on the west side.

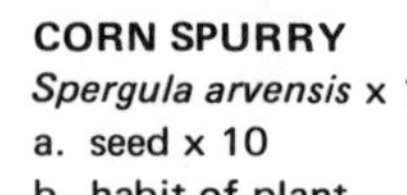

CORN SPURRY
Spergula arvensis x 1
a. seed x 10
b. habit of plant

KNAWEL or GERMAN KNOTWORT

Scleranthus annuus L.

Low annual, with generally clustered branching stems from a long taproot; leaves narrow, pointed, borne in pairs, their bases united; flowers small, greenish, borne singly in angles between branches, also at apex of the stem and in axils of the upper leaves; fruit 1-seeded, enclosed within the woody calyx, and this entire structure at maturity falling from the plant.

Introduced from Europe, this plant is becoming a troublesome weed of waste areas, lawns and gardens in several scattered localities in the Pacific Northwest. It seeds readily and abundantly, and thus spreads rapidly from initial infested areas.

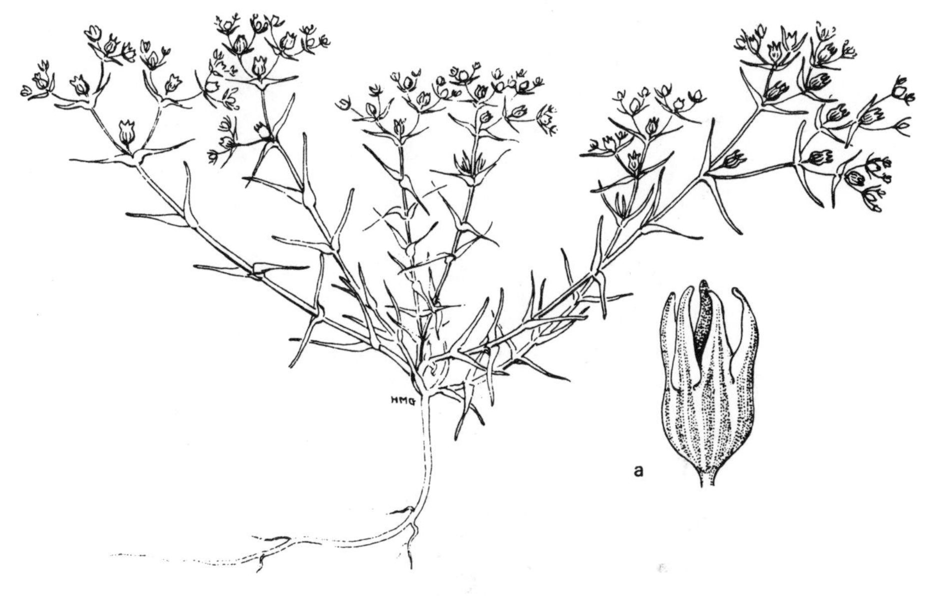

KNAWEL
Scleranthus annuus x 2/3
a. calyx enclosing the fruit x 8

JAGGED CHICKWEED or UMBRELLA SPURRY

Holosteum umbellatum L.

Annual; 1 1/2 to 12 inches tall, with erect succulent branching stems and foliage, the whole plant except the basal leaves more or less covered by soft gland-tipped hairs; basal leaves stalked, stem leaves paired, without stalks, blades oval or elliptical; flowers chickweed-like but borne in umbrella-rayed clusters, these generally long-stalked or, in dwarf plants, with stalks very short; petals 5, white, thin, ragged at the apex.

A native of Eurasia, this species has become established east of the Cascades, especially along the Columbia and Snake rivers, sometimes occurring as a grainfield weed. The plants of early spring are often dwarfed, blossoming when only an inch or two high; plants developed later in the season grow taller.

JAGGED CHICKWEED
Holosteum umbellatum x 1/2
a. open fruit enclosed in calyx x 3
b. seed x 30

MOUSEEAR CHICKWEED

Cerastium vulgatum L.

Perennial; stems generally much branched at base, the stems often spreading, bent, and rooting at the lower joints, the flowering tips erect; entire plant somewhat sticky-hairy; leaves generally narrow; flowers white, rather inconspicuous, borne in small clusters at ends of short branches, each flower with a slender stalk; fruit elongated, narrow, transparent, 10-toothed at the apex; seeds many, very minute, orange-brown, under a lens seen to be wedge-shaped, with wart-like projections.

This Eurasian weed has become a pest in lawns and golf greens, producing unattractive gray-green patches which are untouched by the mower. It also occurs in cultivated fields and gardens where, however, its control is less difficult. West of the Cascades it is found blooming from early spring until late fall.

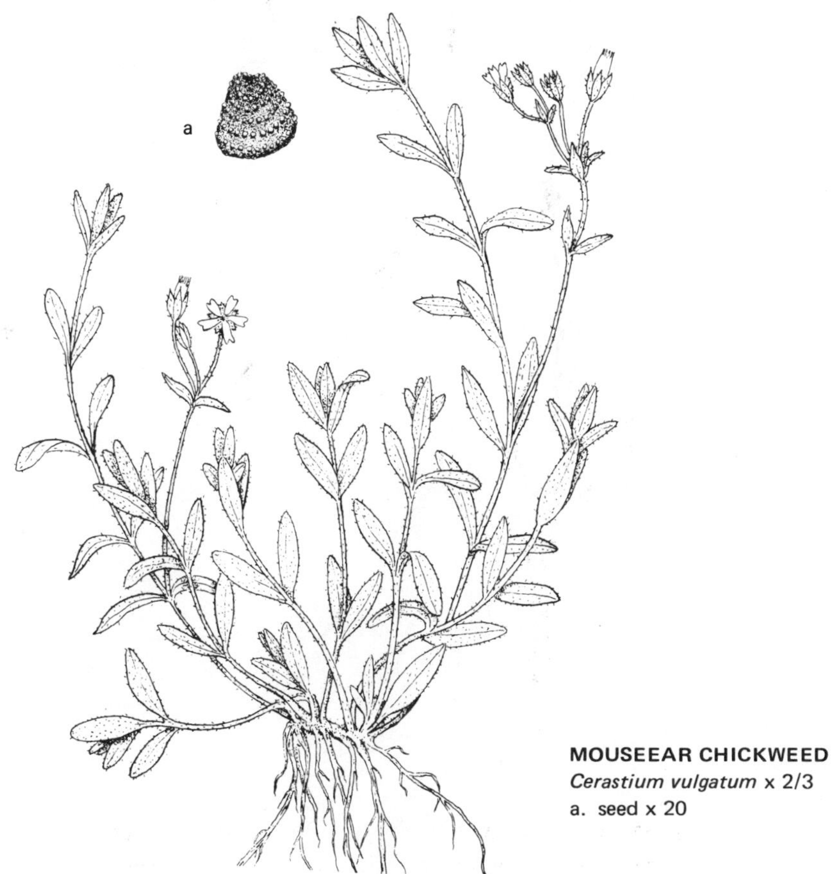

MOUSEEAR CHICKWEED
Cerastium vulgatum x 2/3
a. seed x 20

a. **STICKY CHICKWEED**
Cerastium viscosum x 1

b. **CHICKWEED**
Stellaria media x 1

STICKY CHICKWEED

Cerastium viscosum L.

Annual; spreading or erect, stems 3 to 15 inches long; leaves paired, the upper leaves generally broadest at the base or sometimes nearly round, sticky-hairy; flowers small, white, very short-stalked; fruit small, narrow, opening only at the apex to allow escape of seeds, the empty fruit whitish, nearly transparent, slightly curved, 10-toothed at the apex.

Introduced from Eurasia. A common lawn and garden weed west of the Cascades, troublesome but less difficult to control than its close relative, Mouseear Chickweed.

CHICKWEED

Stellaria media (L.) Cyr.

Annual or winter annual or in favorable climates sometimes becoming perennial; stems 3 to 18 inches long, branching, more or less prostrate, rooting at the lower nodes, or weakly erect, often becoming tangled masses, each stem bearing a vertical line of minute hairs; leaves paired, ovate, the lower stalked, the upper nearly stalkless; flowers small, white, somewhat star-shaped, with 5 sepals, 5 petals appearing as 10 because deeply notched, stamens 3 to 7, the pollen-bearing tips reddish orange; fruit splitting to release numerous, circular, pebbled, dark brown seeds.

Chickweed, introduced long ago from Eurasia, is now abundant wherever conditions are favorable for its growth. This, with Mustard, is one of the earliest weeds in the spring on the Pacific slope but blossoming plants may be seen here almost any month of the year. On a sunny day in early spring, when chickweed forms extensive ground covers over the as yet uncultivated orchards and gardens, its fragrance is distinctly carnation-like and it is an early attraction for bees. Small birds and various species of wild fowl relish its succulent herbage as well as its seeds. In gardens and strawberry fields, however, it is a troublesome weed, its seeds continuing to germinate as long as conditions remain favorable for growth.

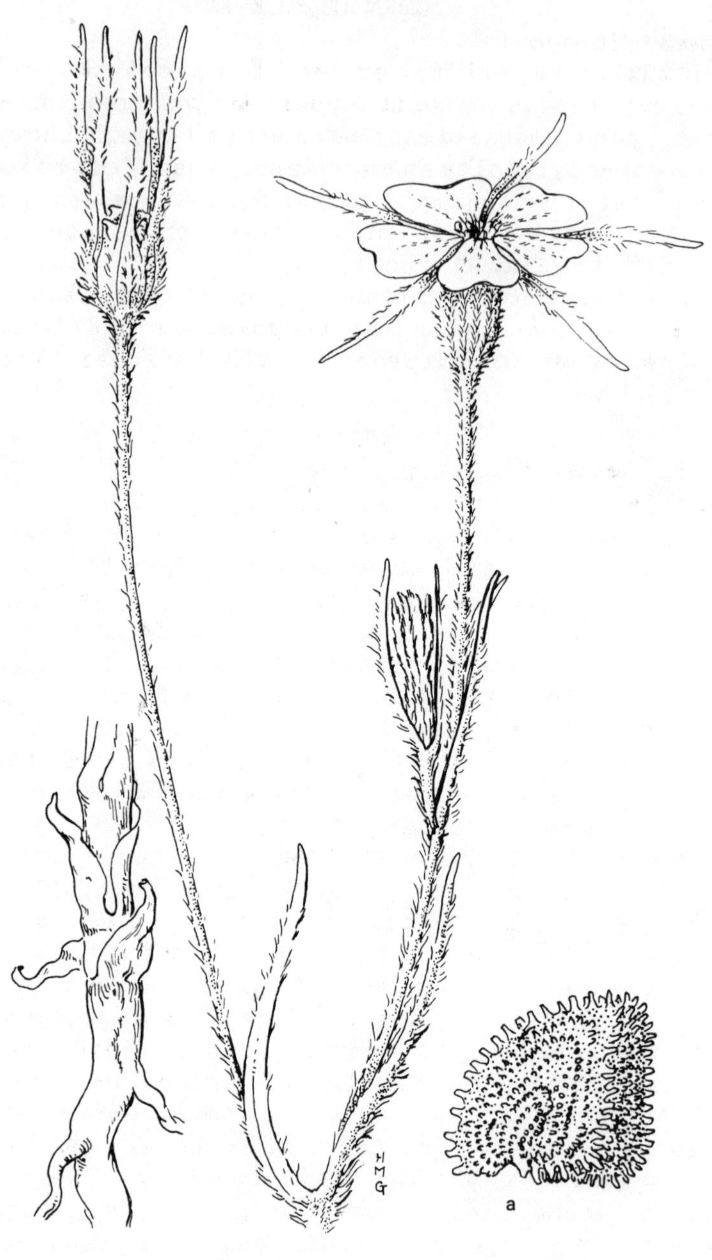

CORN COCKLE
Agrostemma githago x 1 1/4
a. seed x 8

CORN COCKLE

Agrostemma githago L.

Annual; stem stiff, erect, 1 to 3 or 4 feet tall, white-hairy; leaves opposite (paired), narrow, hairy, 2 to 5 inches long, stalkless; flowers large, showy, rose-purple, long-stalked; sepals hairy, fused below to form a 10-ribbed tube, the tips long, narrow, extending beyond the petals; petals broad above, long-clawed below, black-dotted near the base; fruit 1/2 inch long, with toothed opening at the apex to allow escape of seeds; seeds many, very dark, pebbled in concentric rows.

A native of Europe, widely distributed in grainfields, waste land and roadsides.

RED SANDSPURRY

Spergularia rubra (L.) J. & C. Presl

Annual or rarely a short-lived perennial; stems mat-forming; leaves appearing to be arranged in whorls, with conspicuous papery stipules; flowers red, pink or rarely white.

Native of Europe; widespread in the United States in waste places, roadsides, gardens and other cultivated areas.

x 2/3

BIRDFOOT BUTTERCUP
Ranunculus orthorhynchus x 2/3
a. fruit x 6
b. petal x 3

Buttercup Family: RANUNCULACEAE

This common plant family contains many ornamental garden species such as Columbine, Delphinium, Anemone, Aconite, Peony, Christmas Rose (*Helleborus*), Hepatica, and Clematis, as well as Buttercup.

Several members are poisonous, particularly species of Larkspur (*Delphinium*) and Aconite (*Aconitum*). Certain introduced species of Buttercup are so acrid and bitter that they blister the mouth parts of animals, and taint the dairy products of cows feeding upon them. Some of the native species, however, provide nourishing forage in early spring and are therefore of value when grass is sparse.

BIRDFOOT BUTTERCUP

Ranunculus orthorhynchus Hook.

Perennial; 1/2 to 2 feet or more tall; leaves and stems often clustered at the base; basal leaves long-stalked, the blades cut into narrow divisions, these again toothed, the stem leaves similarly cut but smaller; flowers 1 to 1 1/2 inches in diameter, petals typically 5, sometimes elongated, yellow, in some plants rusty or purplish on the back; fruit, including the beak, nearly 1/3 inch long; beak long, narrow, straight or nearly so.

Typically this species occurs principally west of the Cascades, but at least 1 variety of it is found east of the mountains. It is confined to wet land and may be abundant in low pastures. In common with many other buttercups, it appears to be avoided by stock when better forage is available.

CREEPING BUTTERCUP

Ranunculus repens L.

Perennial; stems hairy, often creeping, rooting at the lower nodes; flowering ends of the stems erect; leaves long-stalked, hairy, the blades lobed or divided into 3 segments, these again toothed; flowers golden yellow, shining, stalked, 3/4 to 1 1/2 inches in diameter; seed-heads consisting of approximately a dozen fruits, the individual fruit about 1/8 inch long, flattened, rounded, with a short backward turned beak.

This European buttercup was perhaps introduced as a garden plant, since it is one of the most showy of all known species. A horticultural variety with tight double flowers likewise is used for low borders and as a ground cover.

From gardens the single-flowered form, which seeds readily, has escaped and is becoming troublesome in low pastures, particularly on the coastal side of the Coast Range. Its rich rank growth and rapid spread under such conditions results in crowding out the natural grasses; and since this species of buttercup is known to be poisonous, its presence is a double menace.

CREEPING BUTTERCUP
Ranunculus repens x 2/3
a. cluster of fruits from a single flower x 4

TESTICULATE or HORNED-HEAD BUTTERCUP

Ranunculus testiculatus Crantz

Low annual; leaves all basal, somewhat woolly, 3- to several-parted or divided, the divisions narrow; flower stalks single or several, 1/2 to 7 inches in height, each bearing a single yellow flower, this followed by an elongated gray-woolly head of horned fruits; each individual fruit very small, one-seeded, woolly, with a stiff slightly curved beak and with 2 sac-like projections at base.

This species from southeastern Europe and adjacent Asia has become abundantly distributed in several areas east of the Cascades; and in the spring, in spite of its small size, is a conspicuous roadside covering. By the time the gray heads are mature, the foliage has completely withered and disappeared. The small size of the plant and its short season render it of little importance as a pest.

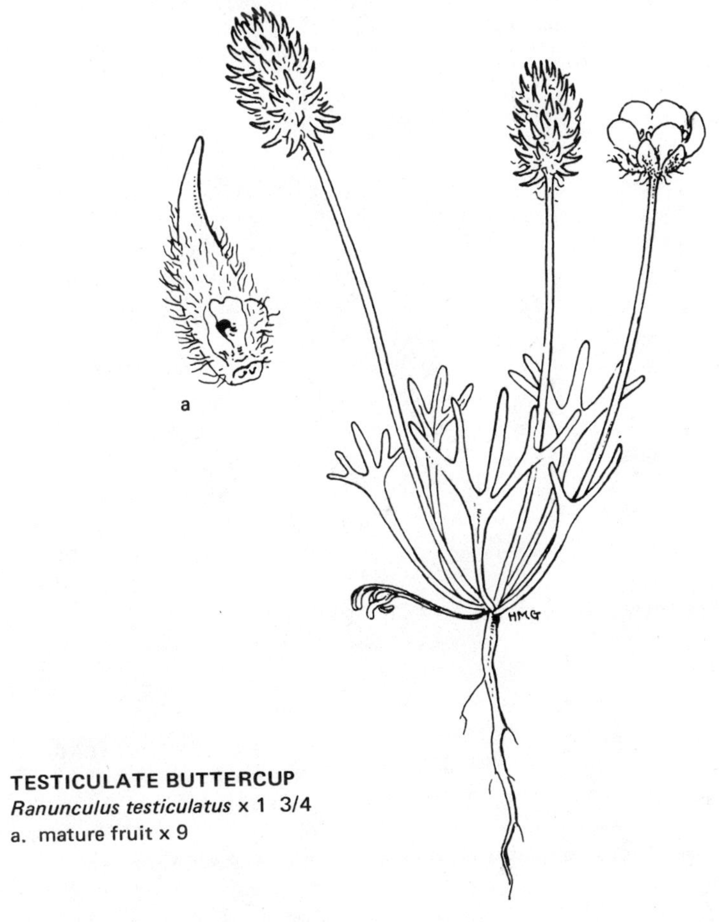

TESTICULATE BUTTERCUP
Ranunculus testiculatus x 1 3/4
a. mature fruit x 9

CORN BUTTERCUP
Ranunculus arvensis
a. plant x 2/3
b. fruit x 4

ROUGHSEED BUTTERCUP
Ranunculus muricatus x 2/3
c. leaf blade x 2/3
d. fruit x 4
e. cluster of fruits from a single flower x 2/3

CELERYLEAF BUTTERCUP
Ranunculus sceleratus
f. fruit x 4
g. cluster of fruits from a single flower x 2
h. leaf blade x 2/3

CORN BUTTERCUP or EUROPEAN FIELD BUTTERCUP

Ranunculus arvensis L.

Annual; stem 3/4 to 1 3/4 feet tall, erect, branched above; basal and lower stem leaves long-stalked, the blade cut into narrow toothed divisions, or sometimes the first leaves scarcely divided; flowers pale yellow, 1/2 to 2/3 inch in diameter; fruits at maturity spreading and conspicuously spiny, the marginal spines longest.

This European weed is sparingly distributed in several counties of western Oregon, also in neighboring states.

ROUGHSEED or SPINY-FRUITED BUTTERCUP

Ranunculus muricatus L.

Annual; 1/6 to 1 foot tall, stems single or clustered, branching, leafy; leaves stalked, the basal leaves sometimes withered at flowering time; leaf blades 3 to 5 lobed, the lobes toothed; flowering stalks shorter or longer than the leaves; flowers yellow, about 1/2 inch in diameter; fruit 1/4 inch or slightly longer, with smooth border, stout curved beak, and generally spiny faces.

This species came from Europe and is established in scattered areas along the Coast and in the Willamette Valley.

CELERYLEAF or CURSED BUTTERCUP

Ranunculus sceleratus L.

Annual; 1/2 to 2 feet tall, stems single or clustered, branched above; lower leaves clustered at base of the stem, long-stalked, the blades deeply cut into 3 to 5 divisions, these again toothed or lobed; flowers small, less than 1/2 inch in diameter, pale yellow; fruits in an elongated head approximatley 1/2 inch long; individual fruit very small, less than 1/16 inch long, smooth, scarcely beaked.

In or on the borders of fresh or alkaline pools or ponds, more common east of the Cascades. The plant when bruised is very irritating to the lips and skin of grazing animals.

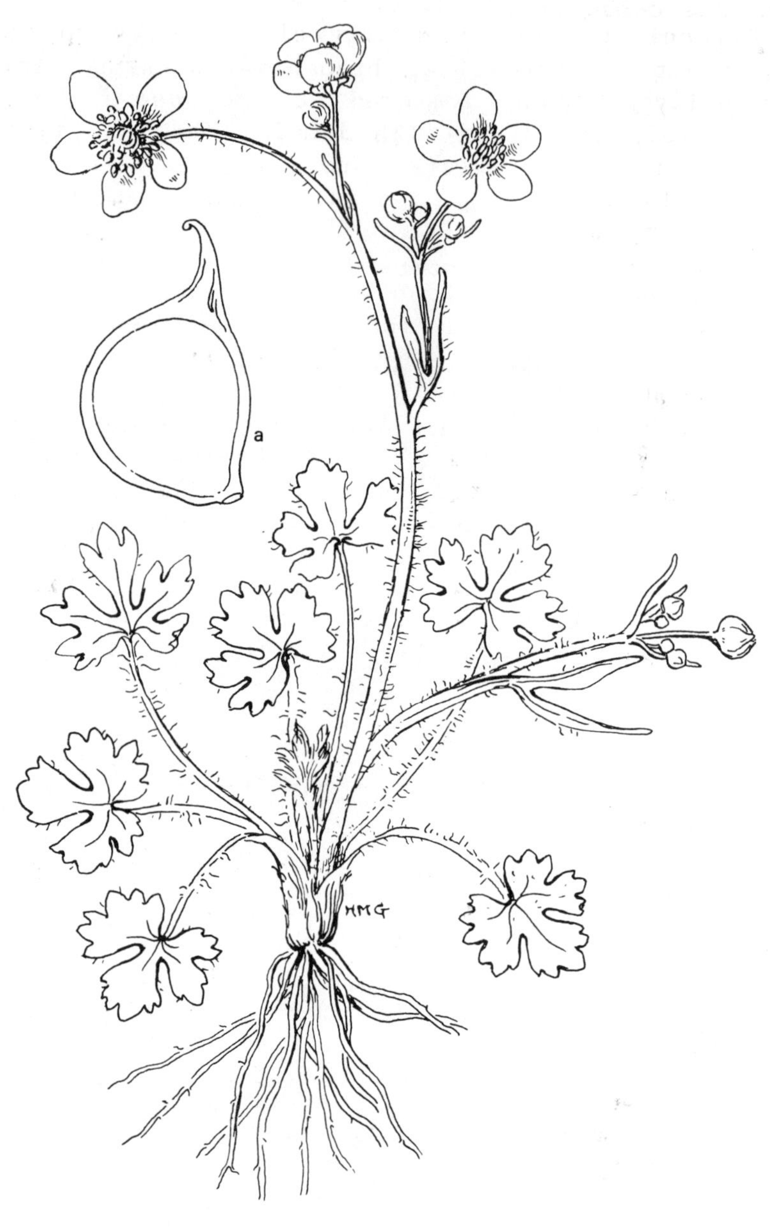

WESTERN FIELD BUTTERCUP
Ranunculus occidentalis x 4/5
a. fruit x 9

WESTERN FIELD BUTTERCUP

Ranunculus occidentalis Nutt.

Perennial; stems generally clustered, 6 to 18 inches tall, erect or spreading, branching, somewhat smooth or often hairy; leaves clustered at the base of the plant, long-stalked, the blade with 3 main lobes, these toothed, hairy; stem leaves with few and narrow divisions; flowers golden yellow, 3/4 to 1 inch in diameter, petals generally 5, rarely more; fruit about 1/8 inch long, with curved beak.

This species is common in fields and on hillsides of the northwestern coast, but less frequent east of the Cascades. Its inclusion in this manual is based on its abundance on western fields and hills in early spring. It is not a noxious weed in cultivated crops, but in pastures it crowds out grasses which would provide better forage, though itself furnishing early feed of considerable value, particularly to sheep.

TALL BUTTERCUP

Ranunculus acris L.

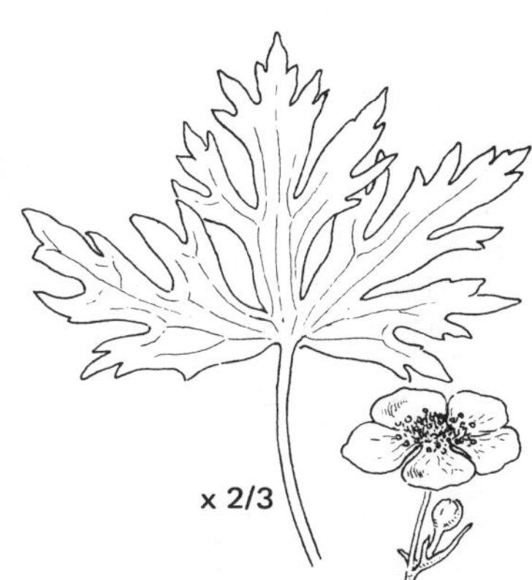

Hairy perennial; stems often reaching 3 feet in height, much branched above; lower leaves deeply 3- to 5-lobed, the lobes deeply cut, upper leaves reduced and consisting of 3 to 4 narrow segments; flowers yellow, an inch or more in diameter.

This species has been introduced from Europe and is now well established throughout most of North America. In the Pacific Northwest it is occasional both east and west of the Cascades; it usually occurs in pastures, and although generally avoided by livestock, it is reported to be poisonous.

TALL LARKSPUR or WOOD LARKSPUR

Delphinium trolliifolium Gray

Perennial; underground structure dark, elongated, woody, branched, bearing fibrous roots; stem 2 to 6 feet tall, stout, leafy; lower leaves long-stalked, the blades generally parted into 5 broad main divisions, these again lobed or deeply toothed; upper leaves shorter-stalked, the lobes narrower, those of the uppermost reduced to 3 or 1; flowers generally many, large, usually deep blue except for 2 small whitish petals; the spur long, stout, more or less curved at the tip, blunt-pointed; fruits smooth, nearly erect.

Common in and along the borders of moist woods and streambanks of the Pacific Slope. This species, which frequently is found in low pastures, is poisonous to stock.

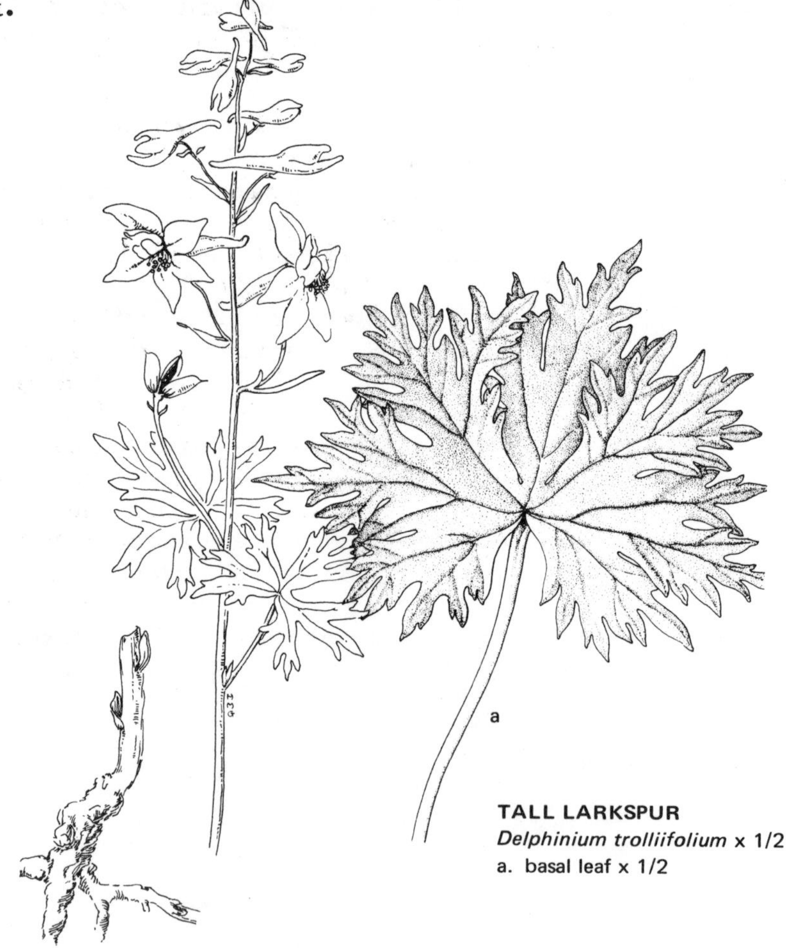

TALL LARKSPUR
Delphinium trolliifolium x 1/2
a. basal leaf x 1/2

DESERT LARKSPUR

Delphinium andersonii Gray

Perennial, with long coarse woody roots; stem 10 inches to 2 1/4 feet tall, single or clustered, simple or branched, often purplish at least below, generally smooth; leaves mostly basal, slender-stalked, the blades 1 1/2 to 2 1/2 inches in diameter, deeply cut into several segments, these again toothed, often gray-green or purple tinged; flowers deep blue, borne in a long open spike, spur somewhat curved; fruits 3 per flower, erect, smooth.

This is a poisonous species found principally on dry hills, open range lands and juniper woodlands east of the Cascades.

x 1/3

PALE LARKSPUR

Delphinium glaucum Wats.

Perennial; stems stout, 2 1/2 to 5 feet tall, whitish, smooth, leafy for the entire length, generally clustered from a large branching woody base; leaves many, slender-stalked, the blades thin, sometimes reaching 8 inches in diameter, divided into 3 main lobes, these again sharply lobed or toothed; flowers blue, in long narrow spikes, the spur nearly straight; fruit shining, nearly erect or somewhat spreading.

This species of Larkspur, which is found mainly in mountain meadows, is poisonous.

x 1/3

FIELD LARKSPUR
Delphinium menziesii x 1/2
a. seed x 10

FIELD LARKSPUR

Delphinium menziesii DC.

Perennial, with a thickened tuber or cluster of tubers from which the stem arises; stem 4 inches to 3 feet tall; leaves borne both at the base and along the stem, the blades 1 to 2 1/2 inches in diameter, deeply lobed and toothed, with narrow segments; flowers showy, in a loose or sometimes compact cluster, dark purplish blue, the spur nearly straight; fruit somewhat spreading.

This is a poisonous species found principally west of the Cascades, especially abundant in fields and on hillsides of the Willamette Valley.

Fifteen to twenty species of Larkspur occur in our area; but since the genus is rather readily recognized, only four species are here included as representative of the group.

CALIFORNIA POPPY
Eschscholtzia californica x 2/3
a. seed x 4

Poppy Family: PAPAVERACEAE

Although this family is well known medicinally and horticulturally, only a few representatives are found outside of gardens in the Pacific Northwest; of these, the California Poppy alone is in sufficient abundance to be considered a weed.

The family is characterized by a milky or watery juice in the stems and leaves, by the presence of generally 4 or 6 petals, and by numerous stamens.

CALIFORNIA POPPY

Eschscholtzia californica Cham.

Perennial but usually flowering the first season; entire plant smooth, grayish green with bitter watery juice; stems 1/2 to 1 1/2 feet tall, generally borne in a dense leafy cluster; leaves several times divided, the final divisions narrow; flowers long-stalked, borne in the leaf axils; petals 4, yellow or orange, pushing off the calyx cap as they open; fruit long, narrow, ribbed, with a flaring collar at the base.

Whether the California Poppy migrated to western Oregon from the south, or is native, is a question; but it probably has been known here from the time of the first white settlers. The plant is a roadside weed, and often invades cultivated fields.

Bleeding-heart Family: FUMARIACEAE

This family is characterized by having irregular shaped flowers with 2 minute sepals and 4 more or less united petals. It is well known for both native and cultivated species of *Dicentra* (Bleeding-heart).

FUMITORY

Fumaria officinalis L.

Annual; stems 1/2 to 3 feet tall, weak, erect or climbing on nearby vegetation; leaves divided into numerous segments; flowers to 5/8 inch long with a sac-like base (the spur), reddish-purple, borne in elongating racemes terminating the stems and from the upper leaf axils; fruit nearly round, but with a small depression at the apex, 1-seeded.

This species is introduced from Europe and is occasinally found as a weed in gardens and cultivated fields, especially wheat.

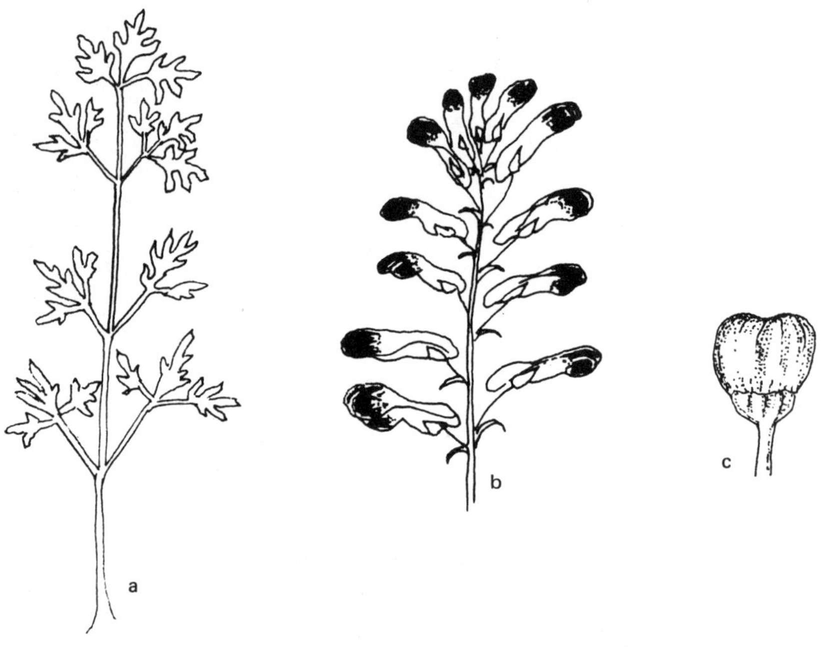

FUMITORY
Fumaria officinalis
a. leaf x 1
b. portion of inflorescence x 1
c. fruit x 5

Mustard Family: CRUCIFERAE

Many garden vegetables and ornamental plants belong to this family. Worth special mention are Cabbage, Cauliflower, Brussels Sprouts, Radish, Kohlrabi, Turnip and Broccoli. Among the ornamentals are Sweet Rocket, Arabis, Sweet Alyssum, and Candytuft.

The flowers typically have 4 petals which spread in the form of a Maltese cross.

VIRGINIA PEPPERWEED

Lepidium virginicum L.

Annual or biennial; stems much branched, up to 2 feet in height; sparsely to densely hairy; basal leaves toothed to deeply cleft; stem leaves reduced upwards, toothed or nearly entire; flowers small, numerous; petals white, fruit nearly round, shallowly notched at the apex.

Several varieties occur in our area differing widely in the lobing of the basal leaves.

A widespread weed of dry open ground throughout the Pacific Northwest and eastward.

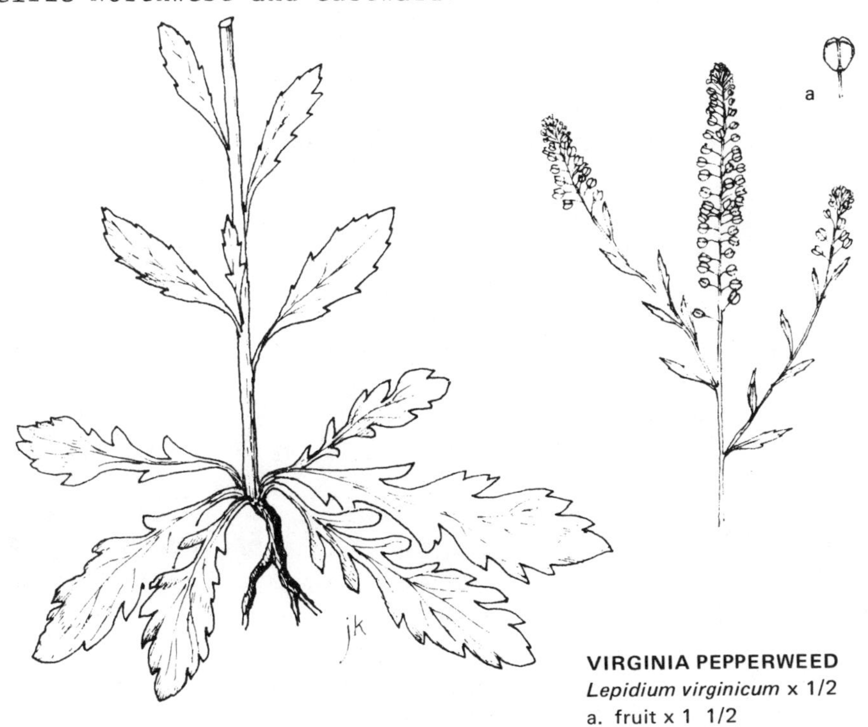

VIRGINIA PEPPERWEED
Lepidium virginicum x 1/2
a. fruit x 1 1/2

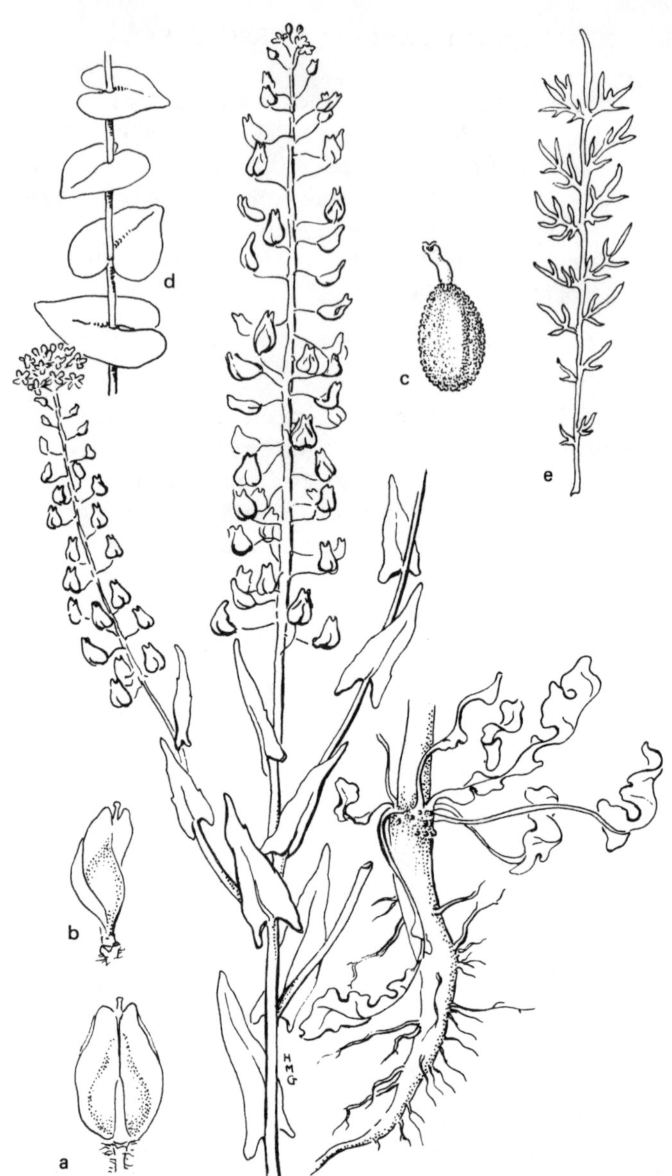

FIELD PEPPERWEED
Lepidium campestre x 2/3
a. ventral view of fruit x 2
b. side view of fruit x 2
c. seed x 6
YELLOWFLOWER PEPPERWEED
Lepidium perfoliatum
d. upper leaves x 2/3
e. lower leaf x 2/3

FIELD PEPPERWEED

Lepidium campestre (L.) Br.

Annual or biennial; stem generally single below (or sometimes several from base of plant), erect, usually branching above, 3/4 to 1 1/2 feet tall, soft-hairy; lower-most leaves long-stalked, the blade with a large terminal lobe and several smaller lateral lobes; upper leaves narrow, clasping the stem, each with 2 ear-like lobes at its base; flowers very small, white; fruit stalked, spreading horizontally, often reaching 1/4 inch in length, slightly curved, with 2 thin terminal wings; seed dark reddish brown, microscopically granular, rounded at the free end, narrowed at the attached end, with 3 somewhat obscure longitudinal ridges.

A European field weed readily propagated by seed, and somewhat established in cultivated fields and range lands. The basal leaves and sometimes the stem leaves also, may wither and completely disappear before maturity, leaving only the bare stalk adorned by numerous fruit.

Another annual species is YELLOWFLOWER PEPPERWEED (*Lepidium perfoliatum* L.). It grows from 1/2 to 1 1/2 feet tall, and is generally widely branched, the stems slender. The basal leaves are several times divided into narrow segments, while the upper leaves are entire, somewhat shield-shaped, and completely encircle the stem. The flowers are yellow, minute, and borne in clusters which elongate as the lower pods mature.

This, too, is a European plant. Occasionally it is encountered on coastward slopes and valleys, but is more common and abundant in fields and dry areas east of the Cascades.

PERENNIAL PEPPERWEED or PEPPERGRASS

Lepidium latifolium L.

Perennial; erect, branching, 1 to 3 feet tall, spreading by a stout creeping underground system sending up new plants at intervals; leaves bright or gray green, shining, somewhat leathery, tapering at the base, the lower leaves stalked; individual flowers very small, white, borne in large dense masses; fruits about 1/10 inch long, flattened, oval, slightly hairy; seed almost microscopic, reddish, minutely roughened.

This species was first reported from Oregon in 1937, but at that time was known from only Umatilla County. It is a potentially serious pest and is becoming established in some of the deeper, richer farmlands. In California, where it has spread widely in several counties, it is believed to have been introduced with beet seed.

With its ability to propagate not only by abundant seed production but also by its stout woody spreading underground system which quickly establishes colonies after the manner of Canada Thistle, this weed is a formidable introduction.

PERENNIAL PEPPERWEED
Lepidium latifolium x 1/2
a. fruits x 6
b. seed x 10

SMALLSEED FALSEFLAX

Camelina microcarpa Andrz. ex DC.

Annual; stem erect, simple or branched, 1 to 3 feet tall, bearing both simple and branched hairs at the base, often smooth above; leaves entire or nearly so, 3/4 to 3 inches long, the upper leaves clasping the stem with basal ear-like lobes, at least the lower leaves hairy; flowers small in elongated, often compound racemes; petals pale yellow to nearly white; fruit less than 1/4 inch long, somewhat inflated, broad above, narrowed to the stalk, the two margins prominent; seeds several per fruit, reddish brown, 1/25 of an inch or less, microscopically granular.

A native of Europe and widely introduced throughout western North America. An occasional weed of grain fields and wasteland particularly east of the Cascades.

LARGESEED FALSEFLAX

Camelina sativa (L.) Crantz

Similar to *Camelina microcarpa* but long hairs not present on lower stems and leaves; fruit larger (averaging about 3/8 of an inch broad and over 1/4 inch long); seeds twice as large as in Smallseed Falseflax and yellowish brown.

Much less common than *Camelina microcarpa* and perhaps not truly established.

SMALLSEED FALSEFLAX
Camelina microcarpa x 1
a. fruits x 2
b. seeds x 50

LARGESEED FALSEFLAX
Camelina sativa
c. fruit x 2

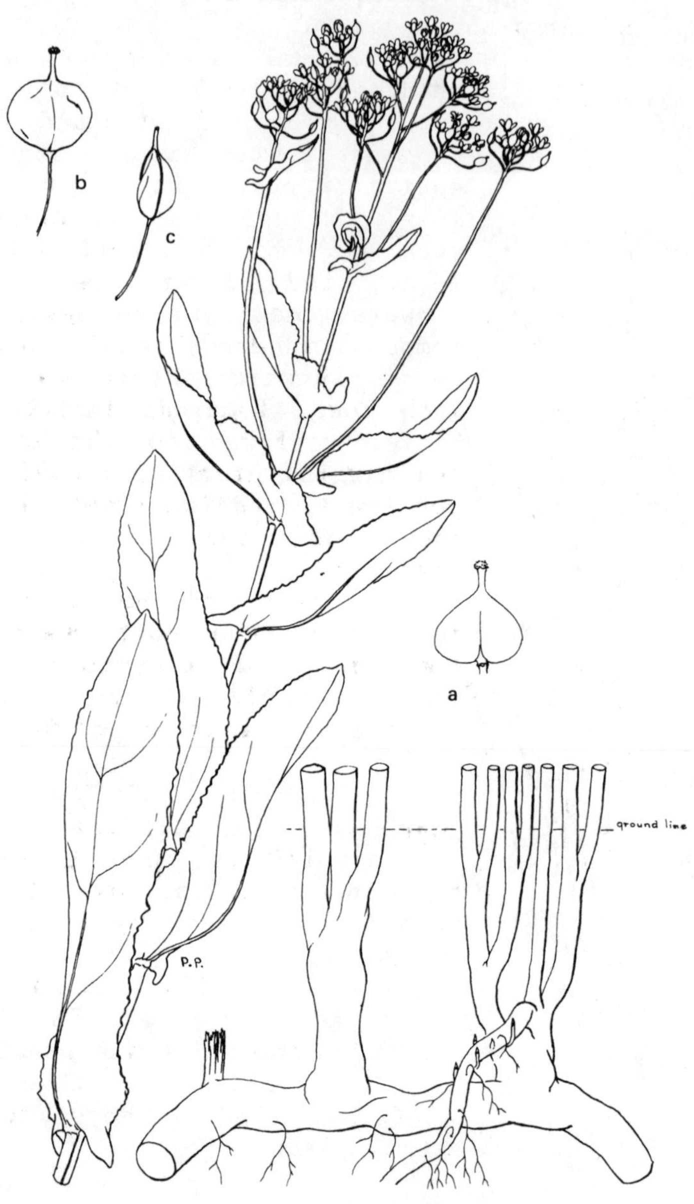

HOARY CRESS
Cardaria draba x 1
a. fruit x 2
Cardaria draba var. *repens*
b-c. fruits x 2

HOARY CRESS or WHITETOP

Cardaria draba (L.) Desv.
(=*Lepidium draba* L.)

Perennial by stout creeping rootstocks; stem 1/2 to 2 feet tall; entire plant minutely hairy, grayish green; leaves generally long-elliptic, the lower leaves stalked, the upper stalkless but with 2 ear-like lobes at the junction with the stem; flowering branches generally numerous, slender, bearing many very small white flowers, these replaced gradually by heart-shaped fruits, each containing 2 compartments separated by a narrow partition, each compartment usually bearing a single reddish brown seed.

Variety *repens* (Schrenk) Schulz. This variety differs from the type form of the species in having rounded instead of heart-shaped fruits, and frequently bears 2 seeds in each compartment.

Pernicious grainfield weeds, both the species and variety are natives of Europe.

HAIRY WHITETOP

Cardaria pubescens (C. A. Mey.) Rol. var. *elongata* Rol.

Similar in size and general appearance, when flowering, to *Cardaria draba*, but with small hairy balloon-like fruits often borne in huge masses; seeds reddish brown, somewhat narrow, slightly curved, rounded at one end, narrowed at the other, with a short white appendage.

This species, like *Cardaria draba*, is a noxious weed in grainfields east of the Cascades. It, too, spreads underground, often forming large colonies. It frequently bears an enormous number of fruits, each of which contains 2 compartments, each compartment capable of producing 1 or 2 seeds. Fortunately, however, in the case of both species the fruits are often sterile, the plant therefore being less prolific than appearances indicate. But sufficient seeds are generally produced to add to the menace of spreading rootstocks, and both species propagate at an alarming rate.

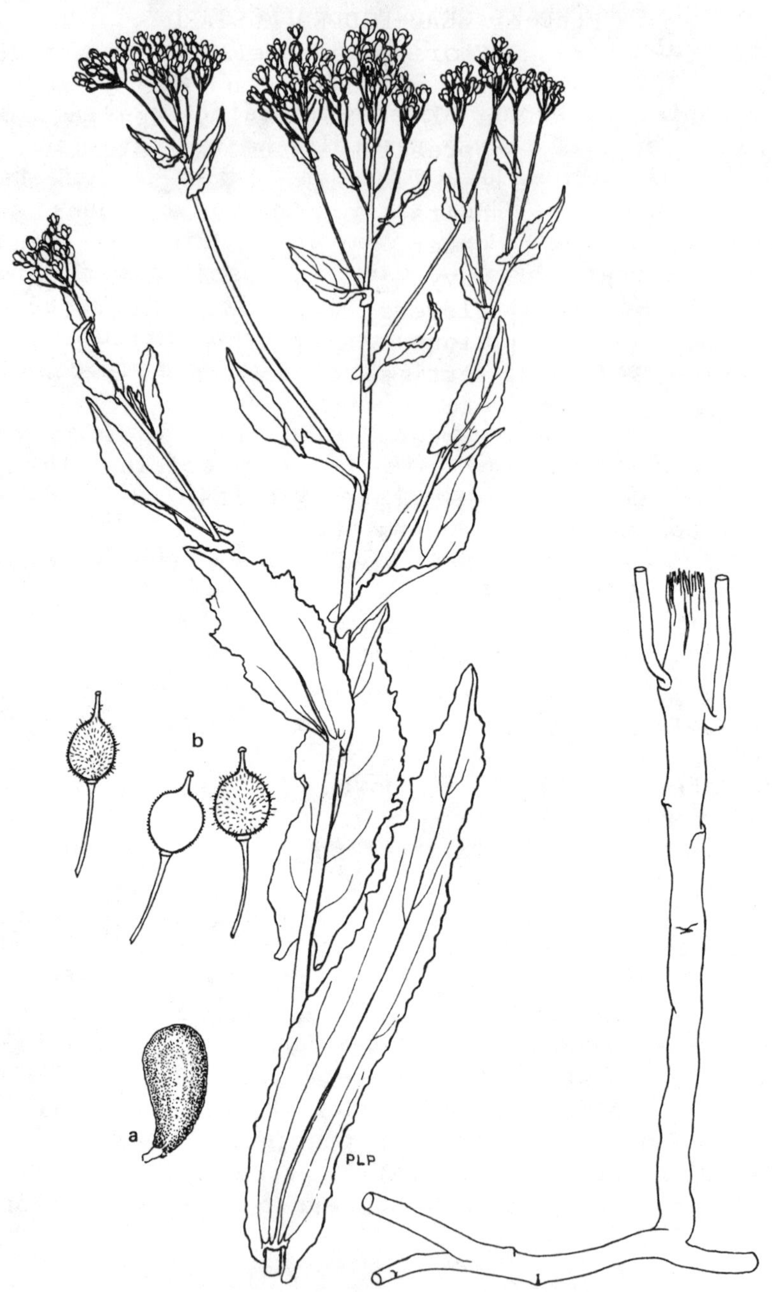

HAIRY WHITETOP
Cardaria pubescens x 1
a. seed x 4
b. fruits x 2

BLUE or BEAD-PODDED MUSTARD

Chorispora tenella R. Br. ex DC.

Annual; plant branching, 1/2 to 1 1/2 feet tall, leafy, somewhat spreading, stem and leaves bearing few to many minute gland-tipped hairs; leaves narrowed at the base, generally stalked, the leaf margins wavy or with a few spreading teeth; flowers small, pale purplish; fruit elongated, broadest at the base, the fertile portion resembling a string of beads, the sterile portion narrowed to a long beak.

Many foreign members of the mustard family have been introduced by accident into America, sometimes probably in imported seed. Blue Mustard is a comparatively recent arrival on the West Coast, but has spread somewhat rapidly in arid regions. Although an annual, it has become an undesirable invader of cultivated land, with an every-ready store of fresh seed supplied from colonies in waste areas.

The fruits do not split longitudinally at maturity, as in the common Mustard, but break apart transversely into a number of 2-seeded sections. These are readily scattered into new areas where they form a nucleus of further colonies. This mustard is a native of southern Russia or southwestern Asia.

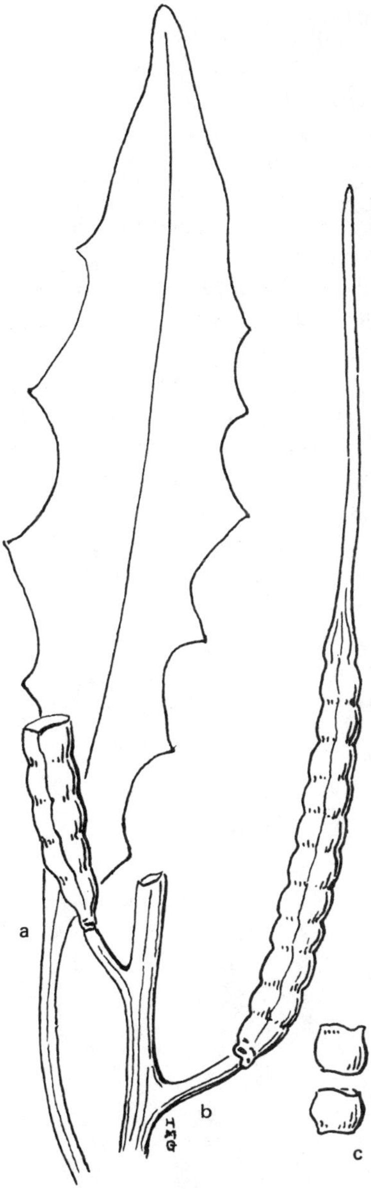

a

b

c

BLUE MUSTARD
Chorispora tenella
a. basal leaf x 2/3
b. fruits x 2/3
c. single-seeded segments of fruit x 2/3

SHEPHERD'S PURSE

Capsella bursa-pastoris (L.) Medic.

Annual; stems slender, erect, 1/4 to 1 1/2 feet tall, one to several from a basal rosette of leaves; lower leaves stalked, generally lobed, the terminal lobe the largest; stem leaves generally toothed, stalkless, clasping the stem and with ear-like lobes at their bases; flowers white, very small, developing in a gradually elongating raceme; fruit small, flattened, wedge-shaped, many seeded, stalked; seed about 1/25 inch long, dark brown.

Introduced long ago from Europe, this is now a common weed of waysides, gardens, farmyards, and waste places. Its common name is derived from the shape of its fruits.

Though an annual and readily hand-pulled, the weed, by its persistence, is a nuisance in gardens, cultivated fields and roadsides.

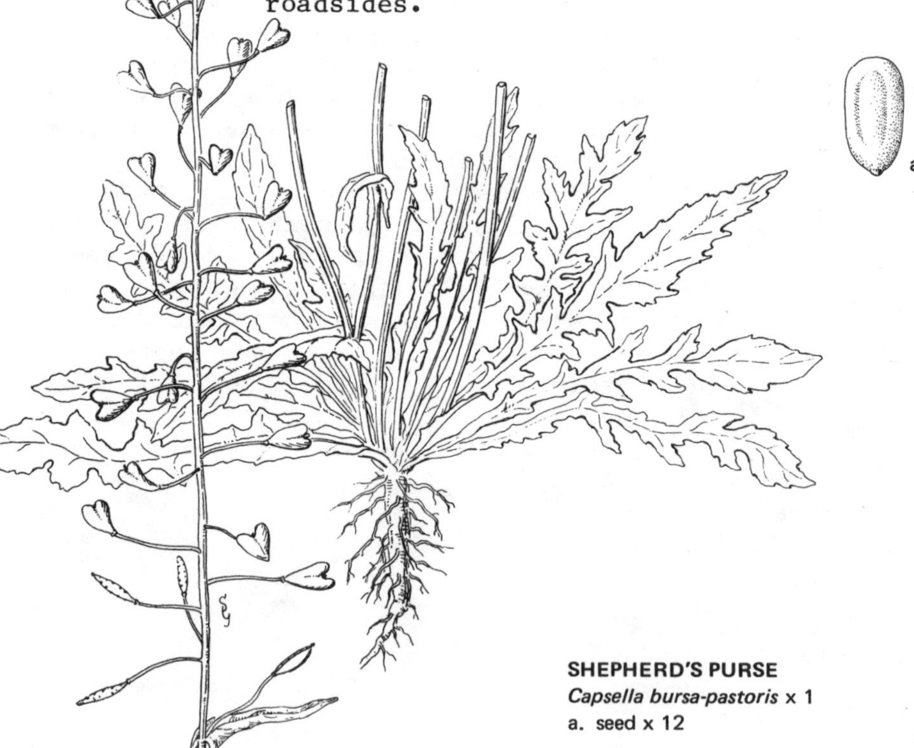

SHEPHERD'S PURSE
Capsella bursa-pastoris x 1
a. seed x 12

FIELD PENNYCRESS

Thlaspi arvense L.

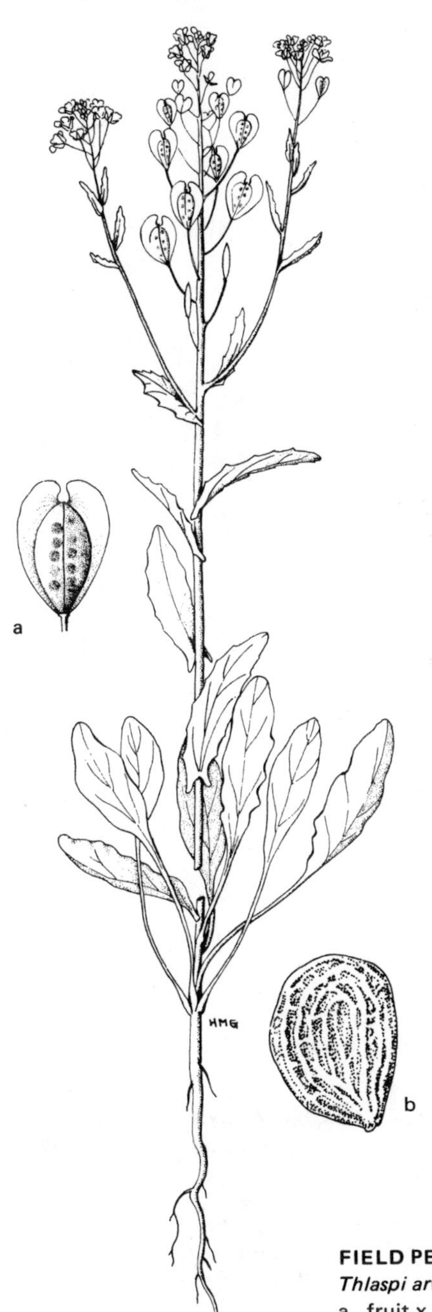

Annual or winter annual; stem erect, unbranched or sometimes slightly branched, 1/2 to 2 feet tall, smooth; lower leaves long-stalked, generally yellowing and withering before flowering time; upper leaves stalkless, closely clasping the stem, with 2 basal ear-like lobes; blades of at least the upper leaves more or less toothed on the margins; flowers very small, white, borne in gradually elongating racemes as the lower fruits mature; fruit generally somewhat longer than wide, flattened, winged, notched at the apex; seed brownish purple, slightly more than 1/16 inch long, somewhat narrower, rounded at the apex, narrower at the base, microscopically ridged, ridges following outline of the seed.

The weed readily propagates by seed and plants only a few inches tall may bear full-sized fruit containing viable seed. This is primarily a weed of grainfields, east but occasionally west of the Cascades in pastures and waste places.

FIELD PENNYCRESS
Thlaspi arvense x 1/2
a. fruit x 1
b. seed x 13

b. **TANSYMUSTARD**
Descurainia pinnata x 1/2

FLIXWEED
Descurainia sophia
c. fruits x 1/2
d. basal leaf x 1/2

a. **HARESEAR MUSTARD**
Conringia orientalis x 1/2

HARESEAR MUSTARD

Conringia orientalis (L.) Dum.

Annual; stem 1 to 2 feet tall, simple or with a few erect upper branches; whole plant smooth; leaves broad, closely clapsing the stem, becoming smaller upward; flowers pale yellow, borne in an elongating cluster; fruits slender, 2 to 4 inches long, somewhat twisted or curved; seeds many, about 1/8 inch long, brown, microscopically roughened.

A European weed not uncommonly found in grainfields, roadsides and waste areas east of the Cascades.

TANSYMUSTARD

Descurainia pinnata (Walt.) Brit.

Annual; stem erect, 1 to 2 feet tall, branching above; leaves divided into small segments, smooth or more often minutely gray-hairy; flowers very small, yellow, long-stalked; seed pods ½ inch or less long, shorter than their stalks.

Tansymustard is a native which is rather widespread east of the Cascades, and attracts attention when occurring in abundance. It seeds prolifically and can be somewhat troublesome in grainfields. It is variable in appearance, several rather distinct varieties being recognized.

FLIXWEED

Descurainia sophia (L.) Webb

Annual; stem 1 to 2 feet tall, often much branched above; leaves several times divided, often fern-like in appearance; flowers cream-colored; fruits slender, 1 inch (or slightly more or less) long, longer than their stalks; seed very small.

This introduced species, though closely related to Tansymustard, can be distinguished by its more finely divided leaves, its paler flowers, and its long slender fruits. With a hand lens, the leaves are seen to be covered by minute star-shaped hairs which sometimes give the plant a gray appearance at first. The leaves often have withered and disappeared before the seed stage. Like Tansymustard, this species seeds abundantly and may become a nuisance in grainfields. Several related species occur, all with similar growth habit.

DYER'S WOAD
Isatis tinctoria x 2/3
a. fruit x 2

DYER'S WOAD

Isatis tinctoria L.

Biennial or perennial; stem reaching 3 feet in height, generally smooth and bluish green; basal leaves long-stalked, widest near the apex; stem leaves stalkless, clasping the stem with ear-like lobes; flowers small, yellow, borne in large terminal clusters; fruits 1/2 to 1 inch long, winged around a thickened center.

This European weed has become a local pest in Idaho and northwestern California. In farming areas it is a noxious weed of pastures and grain fields, where it spreads both by seeds and from the underground system. It has now been reported from both sides of the Cascades in Oregon.

BUSHY WALLFLOWER

Erysimum repandum L.

Grayish annual; stem 6 to 20 inches tall; leaves lance-shaped, up to 6 inches long, usually more or less irregularly toothed; flowers pale yellow; fruit 4-angled, 2 to 4 inches long, somewhat constricted between the seeds.

Waste places and cultivated fields east of the Cascades.

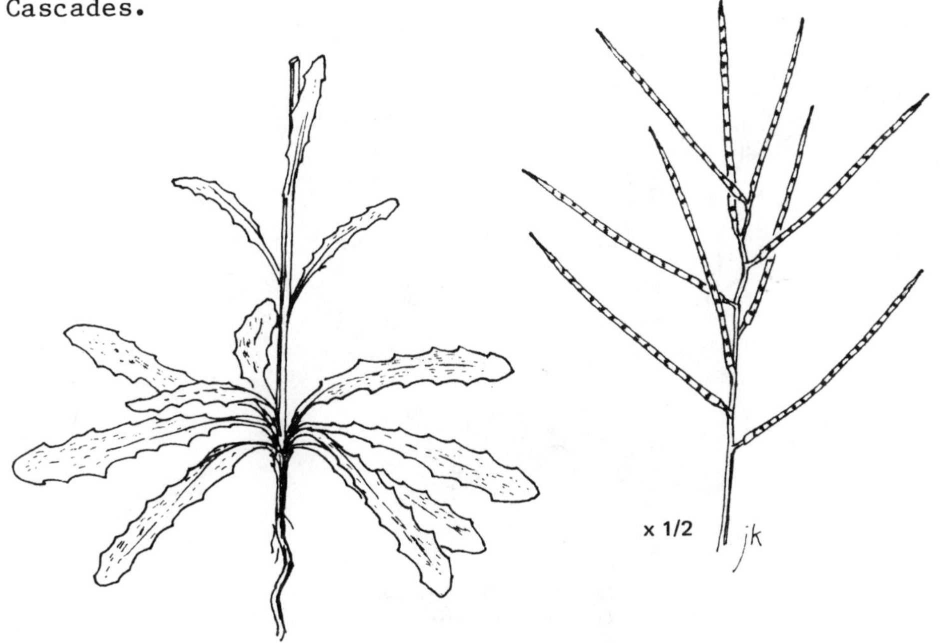

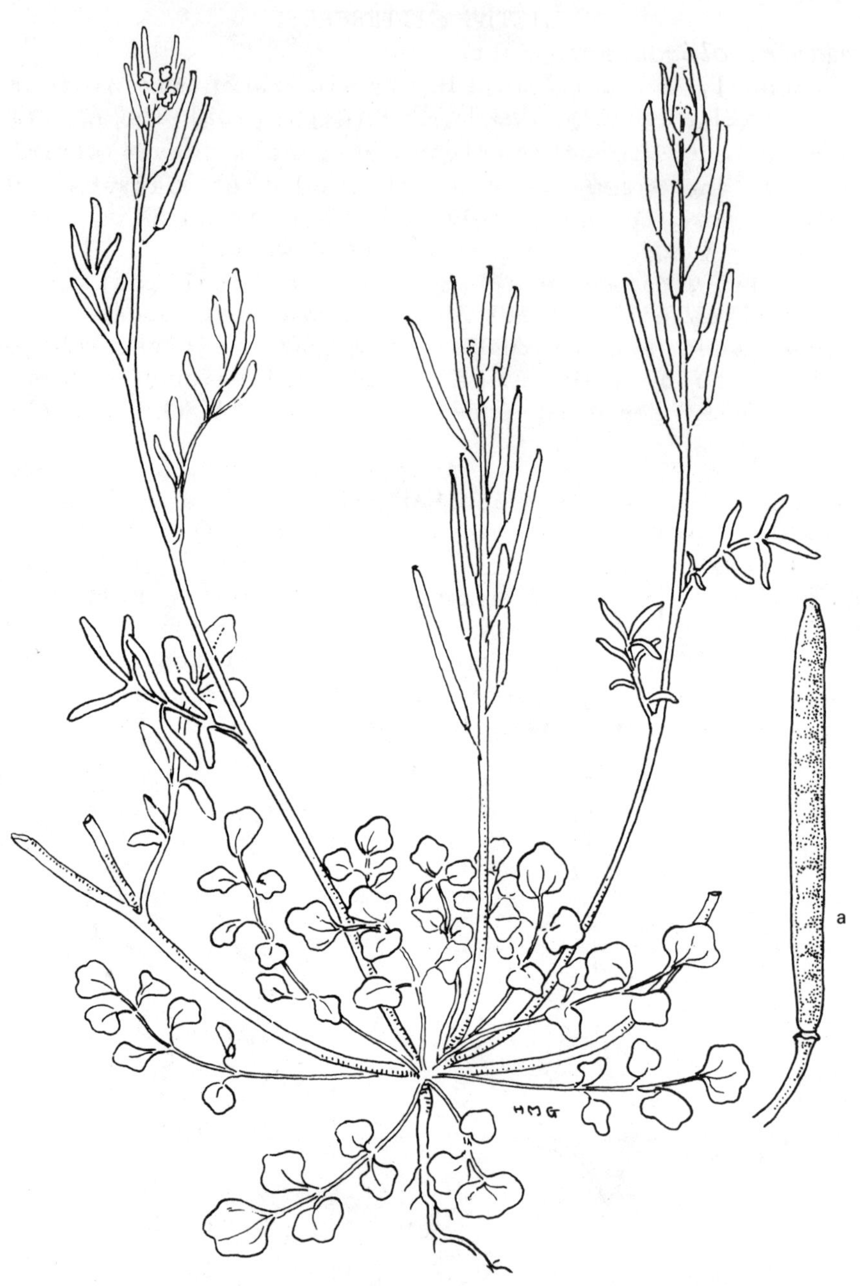

LITTLE BITTERCRESS
Cardamine oligosperma x 2/3
a. fruit x 2 1/2

LITTLE BITTERCRESS

Cardamine oligosperma Nutt.

Annual, winter annual or biennial; stems 2 to 12 inches tall, usually freely branched; basal leaves forming a rosette, each leaf divided into 3 to 11 leaflets, the leaflets generally nearly round, often several-lobed, short-stalked, bright green, the terminal leaflet commonly larger than the lateral ones; leaflets of the upper stem leaves narrower; flowers very small, white, followed by slender fruits 3/4 of an inch long, these stalked, 12- to 24-seeded, location of the seeds externally visible, fruit splitting by two upcurling valves to allow escape of the seeds; seed about 1/20 inch long, flattened, narrowly winged and microscopically pebbled.

Little Bittercress is a native of the Pacific Northwest, and is more common west of the Cascades. It can scarcely be considered a noxious weed. However, its prompt appearance in gardens following seedbed preparation, its lush growth in rainy weather, and its early seeding make it a nuisance. It produces an abundance of seed which germinates quickly, thus perpetuating a succession of the weed until, as the season advances, conditions may become too dry. It generally is accompanied by Chickweed and Shepherd's Purse in newly cultivated moist soil and though, like these, is easily controlled temporarily by hand pulling or cultivating, its persistence, like theirs, necessitates equal persistence on the part of the gardener.

Cardamine oligosperma is an extremely variable species and is not always easy to distinguish from *Cardamine pensylvanica* Muhl. (Pennsylvania Bittercress). *Cardamine pensylvanica* has narrower leaflets and 20 to 40 seeds per fruit. It occurs on both sides of the Cascades, but is more abundant on the east side.

TUMBLE MUSTARD
Sisymbrium altissimum x 2/3
a. fruit x 2/3
b. seeds x 8

TUMBLE MUSTARD or JIM HILL MUSTARD

Sisymbrium altissimum L.

Annual, winter annual, or biennial; 1 to 5 feet tall, bushy-branching, hairy or nearly smooth, first leaves deeply divided into many broad lobes or leaflets; flowers pale yellow, numerous, in lengthening racemes terminating the branches; fruits slender, 2 1/2 to 4 inches long, spreading, with short thick stalks; seed yellowish with dark brown tip.

This weed had its origin in Europe but has long been known in this country. It occurs on both sides of the Cascades, but is in far greater abundance east of the mountains. Like Russian Thistle it may become a "tumble weed" in autumn, breaking off near the ground surface at maturity, assuming a somewhat ball-like form and rolling before the wind, thus scattering its seed.

HEDGE MUSTARD

Sisymbrium officinale (L.) Scop.

Annual with stiff erect stems 1 to 5 feet tall, widely and sparsely branching; lowest leaves clustered at base of the stem, deeply lobed, the terminal lobe largest, lateral lobes generally somewhat pointed and turned bakward, stem leaves few; flowers minute, yellow; fruit 1/2 inch long, pressed closely to the stem; seed 1/16 inch long, brownish, indistinctly pitted.

A native of Europe and now a very common weed of roadsides, farmyards and wasteland. Its large size, sparse and ungainly branching and appressed fruits, distinguish it from other species of *Sisymbrium* and related plants. It is more abundant west than east of the Cascades, and is well known about abandoned farm buildings. It seeds readily and tends to spread to gardens and other cultivated areas. Like Plantain and Mayweed, it has accompanied the white man on his meanderings, and now is known on every continent and in many islands of the larger oceans.

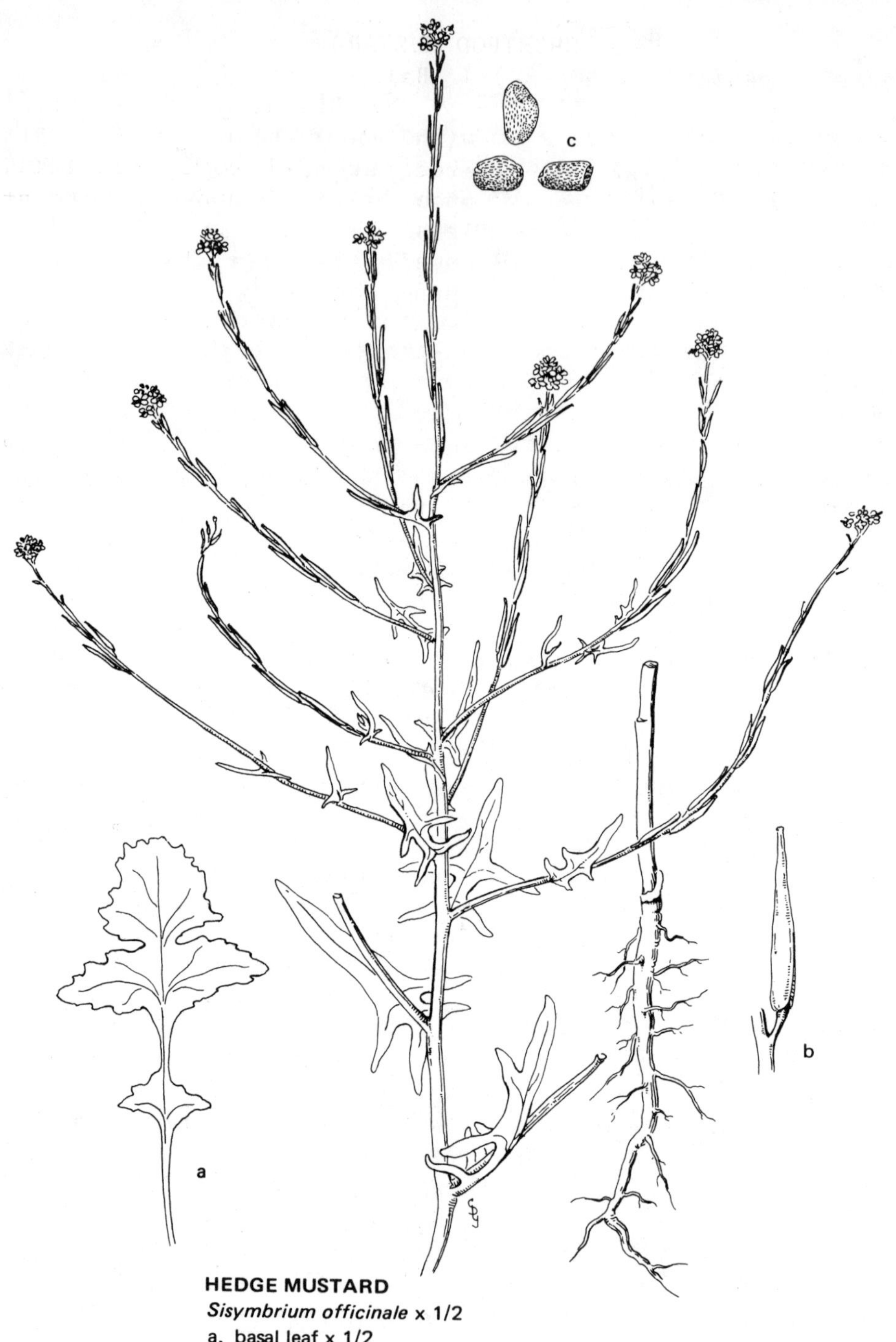

HEDGE MUSTARD
Sisymbrium officinale x 1/2
a. basal leaf x 1/2
b. fruit x 2 1/2
c. seeds x 5

SHORTPOD MUSTARD

Brassica geniculata (Desf.) J. Ball

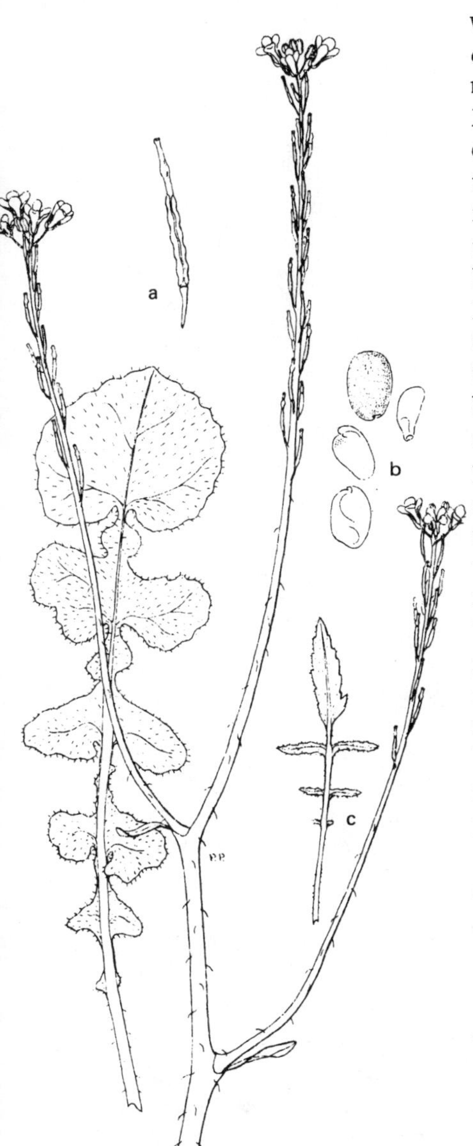

SHORTPOD MUSTARD
Brassica geniculata x 4/5
a. fruit x 2
b. seeds x 10
c. upper stem leaf x 4/5

Biennial or perennial with woody taproot and stiff erect stem, 1 to 3 feet tall, much branched above; stem, at least below, more or less covered by stiff downward-turned hairs; first leaves forming a basal rosette, but these often withered or gone by seed-bearing time, lower leaves raggedly lobed, the terminal lobe much larger than the lateral lobes, rough-hairy, upper leaves small, toothed; flowers small, pale yellow, borne in clusters elongating as the fruits mature; fruits pressed closely to the stem, each about 1/2 inch long, short-stalked, beaked, several-seeded, generally with an extra seed in the beak, fruits at maturity splitting and breaking away, each leaving its thick stalk and thin papery partition; seeds minute (about 1/16 inch long), somewhat variable in shape, generally oblong, netted over the surface.

This species has become established in the Sacramento Valley of California and has spread northward into western Oregon. It is reported as a pest in alfalfa fields.

WILD MUSTARD or CHARLOCK

Brassica kaber (DC.) Wheeler
(=*Brassica arvensis* Rabenh.)

Annual or winter annual; 1/2 to 3 feet tall (rarely considerably taller), somewhat rough-hairy, erect, branching above; lower leaves generally deeply lobed or toothed, the terminal lobe largest, upper leaves generally merely toothed, broadest near the base, short-stalked or stalkless but not clasping the stem; flowers yellow, borne in elongating racemes as the lower fruits ripen; mature fruit 1 to 2 inches long, dark, shining, ridged, more or less distorted by the seeds, beaked, occasionally containing 1 seed in the beak as in Shortpod Mustard (*Brassica geniculata*); seeds globose, usually black, netted or smooth.

Native of Europe; raised for the oil obtained from the seeds, this species often escapes and becomes weedy.

INDIAN MUSTARD

Brassica juncea (L.) Coss.

Annual; 1 1/2 to 4 feet tall, nearly smooth, bluish green; lower leaves with a large terminal lobe and several very small lateral lobes; upper leaves oblong, entire-margined or toothed, short-stalked or stalkless, but not clasping the stem; fruits 1 1/2 to 2 inches long, on slender spreading stalks, each stalk about 1/2 inch long, beak slender, approximately 1/4 the length of the entire fruit; seeds red-brown with a netted surface.

This is an Asiatic species which has found its way to America and has become a noxious weed in certain Middle Western States. Thus far it has been found only occasionally in fields and on roadsides of the North Pacific States, but its further spread should be discouraged.

WHITE MUSTARD

Brassica hirta Moench

Annual or winter annual; 1 to 3 1/2 feet tall, erect, smooth or slightly stiff-hairy; leaves generally all lobed or divided, the upper smaller; flowers yellow; fruit 1 to 1 1/2 inches long, spreading, densely short-hairy, distended bead-like over the seeds, the valves flattened and broad at the base; seeds pale yellow.

White Mustard is often cultivated and occasionally escapes; it is frequently reported from cultivated fields, though it does not occur in abundance. It somewhat resembles *Brassica kaber*, but differs in a longer-beaked fruit and fewer seeds (generally 2 or 3 rather than 5 or 6).

WILD MUSTARD
Brassica kaber
a. upper portion of plant x 2/3
b. basal portion of plant x 2/3
c. seed x 6

INDIAN MUSTARD
Brassica juncea
d. basal leaf x 2/3
e. fruit x 2/3

WHITE MUSTARD
Brassica hirta
f. basal leaf x 2/3
g. fruit x 1

WILD TURNIP
Brassica campestris x 2/3
a. fruit x 1 1/3
b. seed x 4

WILD TURNIP or MUSTARD

Brassica campestris L.

Biennial or winter annual; stem blue-green, erect, 1 to 4 feet tall, simple or branching; lower leaves from a few inches to a foot or more long, divided into a large terminal lobe and several smaller lateral lobes, smooth or bearing a few stiff hairs, these leaves often dying and disappearing before maturity of flower and seed; upper stem leaves not lobed, but broadened at the base and clasping the stem, buds and flowers yellow, borne at first in a flat-topped cluster, this elongating as the flowers open successively upward; fruit 1 1/2 to 2 1/2 inches long, slender, narrowing to a conspicuous beak, seeds nearly black, globose.

West of the Cascades and in various areas east of these mountains this is the most common of our wild mustards. It is native of Europe but was introduced long ago into America, probably in impure seed. The foliage and buds are edible and furnish a readily available source of palatable "greens" in the spring wherever the plant is abundant. Mustard is also a honey plant, and glands in the flower secrete nectar which is widely used by bees.

In areas of Oregon in which the winters are mild, there is scarcely a season when Mustard flowers may not be seen. Seeds of the current season often germinate in the fall, with the plants reaching maturity the next spring. Over-wintering seeds will germinate in the spring, and build up a strong vegetative growth during the growing season, with large basal leaves and a thick root stored with food, by the beginning of winter. This stage, with flower buds already formed, will blossom with the first encouraging days, even in December.

The fruits open explosively by the splitting off of two valves, and the seeds may be hurled some distance from the parent plant, a large specimen of which may produce several thousand seeds.

BLACK MUSTARD
Brassica nigra
a. upper portion of plant x 2/3
b. fruit x 1 3/4

WINTERCRESS
Barbarea orthoceras
c. upper portion of plant x 2/3
d. basal leaf x 1

BLACK MUSTARD

Brassica nigra (L.) Koch

Annual; stem erect, more or less branched, 2 to 5 feet tall, somewhat stiff-hairy below but generally smooth above; leaves stalked, the lower deeply lobed, the upper toothed, smooth or with a few stiff short scattered hairs; flowers small, borne in gradually elongating clusters; fruits stalked, erect, slender.

As is true of most of our other mustards, this species has been introduced from Europe, and it is becoming a common noxious weed of cultivated fields, roadsides, and waste areas both east and west of the Cascades. This species and Short-pod Mustard, which is of more recent introduction, occur in similar situations and somewhat resemble each other, particularly in the small erect fruits. The two may be distinguished, however, by the fact that Short—pod Mustard is more conspicuously bristly, and its fruit generally bears a seed in the beak (see illustation), giving it a characteristic knotted appearance.

WINTERCRESS

Barbarea orthoceras Ledeb.

Biennial, perennial or sometimes a winter annual; plant dark green, erect, simple or branching, 1/2 to 2 1/2 feet tall; lowermost leaves generally rounded and with a few small lobes or leaflets, these leaves usually gone by flowering time, other leaves generally cut into numerous lobes or leaflets, the terminal leaflet or lobe rounded or somewhat elongate, usually much larger than the small lateral structures; petals 3/16 inch long or less, yellow, flowers at first borne in clusters at the tips of the stems, but these elongating as the flowers mature; fruits slender, 3/4 to 2 1/4 inches long, nearly erect to spreading; seed finely pitted.

This is a mustard-like plant commonly seen throughout our area in the early spring. It differs from *Brassica campestris* by darker green foliage, smaller flowers and divided stem leaves.

Two similar species of *Barbarea* are occasional in the Pacific Northwest. *Barbarea vulgaris* R. Br. (Yellow Rocket) is found in the Puget Sound area and occasionally in Oregon. It differs from *Barbarea orthoceras* in possessing a beak-like style on the fruit and petals 1/8 inch long or more. *Barbarea verna* (Mill.) Asch. (Early Wintercress) is an occasional escape from cultivation. It has much divided basal leaves and larger flowers than *Barbarea orthoceras*.

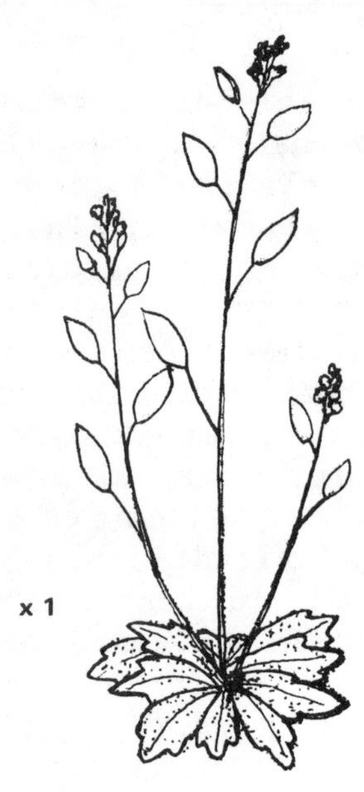

WHITLOWWORT or WHITLOWGRASS

Draba verna L.

Annual; leaves all basal, 1/4 to 1 inch long, entire or toothed near the apex; flowering stalks 1 to 6 inches tall; flowers several, white, petals cleft to the middle; fruit oval, the 2 valves falling away early in fruit, leaving the partition membrane; seeds minute, light reddish-brown, flattened and notched at the base.

Common in a variety of habitats (from orchards to desert areas) in early spring, making conspicuous white patches when in bloom in spite of the diminutive size of the individual plant. This species occurs throughout most of North America and in Eurasia.

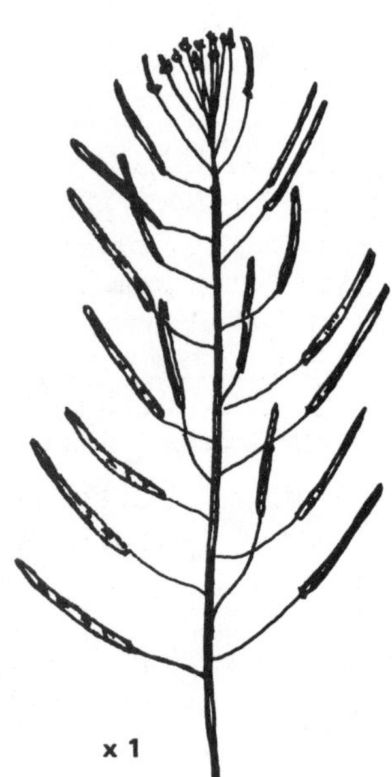

MOUSEEAR CRESS

Arabidopsis thaliana (L.) Heynh.

Annual to 16 inches in height; stems hairy near the base; basal rosette of leaves present, these remotely toothed and bearing both simple and forked hairs; stems with few leaves; flowers small, numerous, white; fruit to 3/4 of an inch long; seeds numerous.

European species now common as a weed in much of the United States, in our area it is common in dry fields, along roadsides and in waste places.

WILD RADISH

Raphanus sativus L.

Annual or biennial, from a long somewhat thickened taproot; stem 1 to 4 feet high, more or less branching, often bearing a few stiff hairs; lower leaves pinnately lobed, the terminal lobe much larger than the lateral; upper leaves smaller; flowers 1/2 to 3/4 inch long, commonly purple, white or yellow with purple veins; fruit broad below, narrow and pointed above, somewhat veined longitudinally, 1- to 3- or rarely 5-seeded, seed about 1/8 inch long, brown or yellowish, covered by a fine network.

A European plant which has become widely established and is a common weed in cultivated fields. Closely related to it is another species also called Wild Radish or Jointed Charlock (*Raphanus raphanistrum* L.), which differs principally in having usually yellow dark-veined or tinged flowers, occasinally white; fruit proportionately narrower, up to 12-seeded and, when dry, appearing like strings of beads.

WILD RADISH
Raphanus sativus x 4/5
a. seed x 4

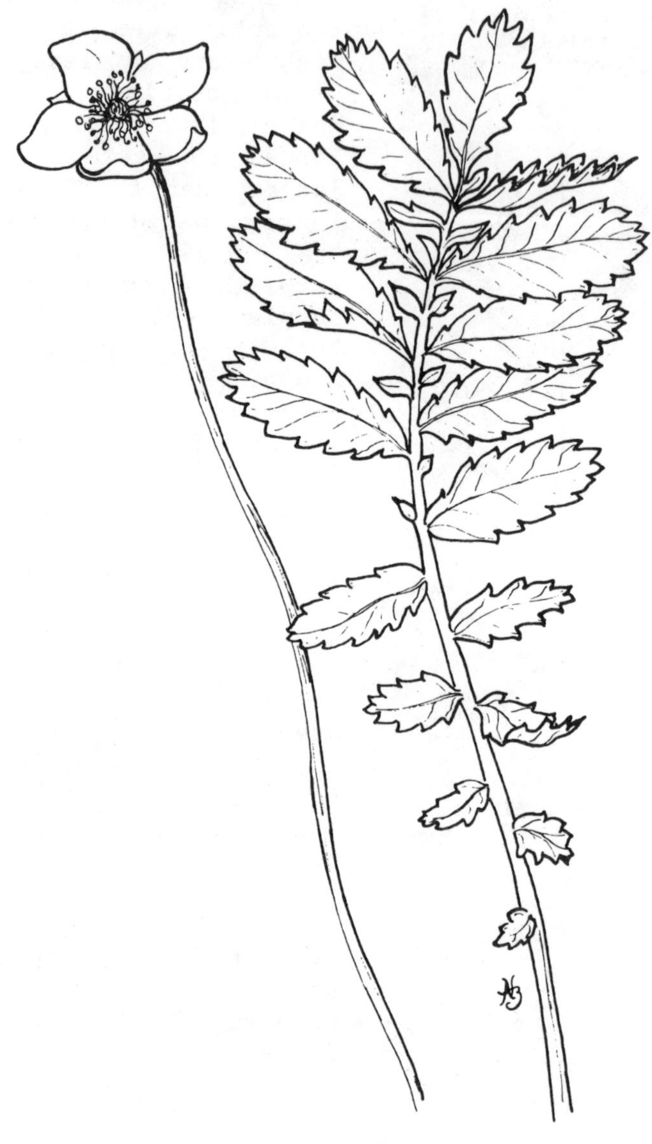

SILVERWEED
Potentilla anserina x 1

Rose Family: ROSACEAE

Although this is one of our best known and most widely distributed plant families, it contains few species which deserve classification as noxious weeds. Curiously, several of the few which has proved sufficiently troublesome to require serious control measures are shrubs rather than herbs.

The family contains many useful and ornamental species. The larger proportion of fruits grown in the North Temerature Zone such as Apple, Pear, Quince, Plum, Peach, Cherry, Raspberry, Blackberry and Strawberry may be found here.

Though often widely varied in appearance, members of the family have certain characters in common. The flower is generally built on the plan of 5, with stamens usually 10 or more, all the parts borne on or within a more or less conspicuous rim-like structure. Certain representatives of the family, as Spiraea, have dry rather than fleshy fruiting parts.

SILVERWEED or CINQUEFOIL

Potentilla anserina L.

A low perennial creeping by strawberry-like stolons; leaves cut into many leaflets, small ones alternating with the larger, leaflets dark green (or rarely silvery) above, silvery-silky beneath; flowers resembling strawberry blossoms in form but golden yellow and each borne singly on its stalk; fruiting heads dry, not fleshy as in strawberry.

This species, with slight variations, is found in the North Temperate Zone of Europe, Asia, and America. It grows in wet places mainly east of the Cascades. In low interior pasture and on irrigated land it tends to become a nuisance by crowding out valuable forage plants. Its habit of spreading by stolons or "runners" gives it an advantage over most of its neighbors, and a single plant may rapidly overrun a large area.

A closely related species, *Potentilla pacifica* Howell (Pacific Silverweed), is essentially coastal. It differs in having nearly smooth fruits as opposed to the ridged or wrinkled fruits of *Potentilla anserina*.

SWEETBRIAR ROSE

Rosa eglanteria L.

Shrub often reaching 10 or more feet in height, the stout clustered pale green stems arching and spreading, and thickly clothed with stiff, sharp, downward-curved prickles; leaves with generally 5 to 7 leaflets, covered beneath by minute glands, the foliage very fragrant; flowers generally more or less clustered, on prickly stalks, petals clear pink with a white base; fruits large, elongated, bright orange-red.

This European species apparently was introduced many years ago along the Pacific Coast as a garden shrub. Readily becoming naturalized, it has spread widely over pasture and wastelands west of the Cascades, and is frequently a companion, along fence rows, of several native species of rose. It is distinguished from them by its arching pale green stems, sweet scent, stout and strongly curved prickles and large elongated fruit.

SWEETBRIAR ROSE *Rosa eglanteria* x 1/3 a. fruits x 1/3

HIMALAYA BLACKBERRY

Rubus procerus Muell.

Shrubs, the stems stout, somewhat erect or trailing, ridged, covered thinly or densely by flattened prickles; leaves with generally 5 leaflets, these rounded, toothed, green above, white-woolly beneath; flowers somewhat showy, white or pale pinkish, borne in generally large clusters, the flower stalks white-woolly; fruit black, shining.

Probably introduced from Europe, and now established in many areas, particularly west of the Cascades, this shrub has become a nuisance in waste places, burned and logged-off areas, along roadsides, and in pastures.

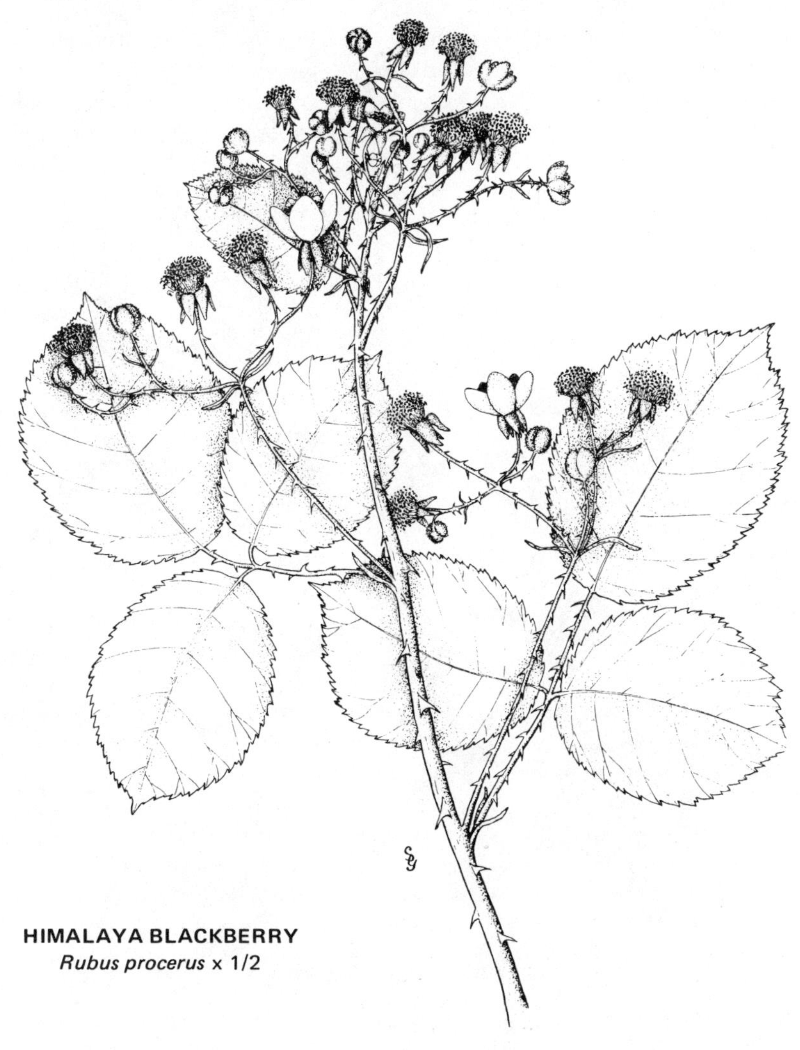

HIMALAYA BLACKBERRY
Rubus procerus x 1/2

EVERGREEN BLACKBERRY
Rubus laciniatus x 1/2
a. lower leaf x 1/4

EVERGREEN BLACKBERRY

Rubus laciniatus Willd.

Shrubs with stout arching often reddish stems armed with stout flattened downward-turning prickles; leaves divided in several leaflets (usually 3 to 7), each leaflet again divided or deeply lobed and sharply toothed, the axis of the leaf likewise armed by sharp recurved prickles; flowers white or pinkish in more or less flat-topped terminal clusters; fruit black at maturity, 3/8 to 5/8 inch in diameter and often reaching more than one inch long.

The origin of this species of Blackberry is probably not known with certainty. It is perhaps an offshoot of a common European species, though some horticulturalists believe that it originated in one of the Pacific Islands. At any rate, it is now widely known in America, having been introduced into the eastern states as an ornamental shrub and having subsequently become extensively distributed. Not only do birds aid in its dispersal, but individual shrubs may, by basal sprouting, eventually form large thickets. Both the stems and leaves are generally evergreen in the milder areas of the Pacific Northwest. This species is very common west and occasionally on the east side of the Cascades.

WESTERN LADY'S-MANTLE

Alchemilla occidentalis Nutt.

Annual; stems 1 to 4 inches long or occasionally longer; leaves short stalked and with large sheathing stipules, blades palmately lobed; flowers borne along most of the length of the stem in clusters in the axils of the leaf stipules, flowers greenish, inconspicuous, petals absent.

This species is found throughout the Pacific States and in the eastern United States. It usually occurs in open areas, often in cultivated fields.

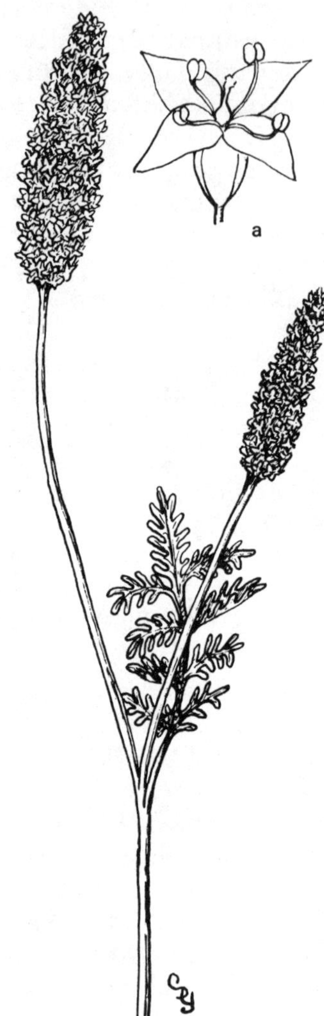

ANNUAL BURNET

Sanguisorba occidentalis Nutt.

Annual; stems 1/2 to 2 feet tall, leafy throughout; leaves 1 to 3 inches long, compound, the leaflets cut into narrow segments; flowers small, greenish, borne in dense spikes 5/8 to 1 1/4 inches long; fruit 1-seeded.

This is a native species occurring in open areas that are moist, at least early in the spring, east of the Cascades in Washington, but on both sides of the Cascades in Oregon, only occasional as a weed in cultivated or disturbed ground.

SMALL BURNET

Sanguisorba minor Scop.

Perennial; stems to 2 feet tall; leaves compound with 7 to 21 leaflets, these nearly circular and coarsely toothed; flowers small, greenish to purple-tinged, borne in compact spikes 3/8 to 3/4 of an inch long; fruit 1-seeded.

Introduced from Europe for possible use as a forage plant, now escaped from cultivation and occasional in waste places throughout our area.

ANNUAL BURNET
Sanguisorba occidentalis x 1 1/2
a. flower x 10

Pea Family; LEGUMINOSAE

This is one of the most widely distributed of plant families and, in general, one of the most easily recognized. With few exceptions it is characterized by compound leaves, irregular flowers having one banner petal, 2 wings, and 2 keel petals, and fruits splitting by 2 valves.

The family includes Bean, Pea, Lentil, Clover, Vetch, Alfalfa, Peanut and many other crop plants. Here, also, belong well-known ornamentals such as Locust, Sweet Pea, Lupine, Wisteria, Acacia, and Laburnum.

SCOTCH BROOM

Cytisus scoparius Link

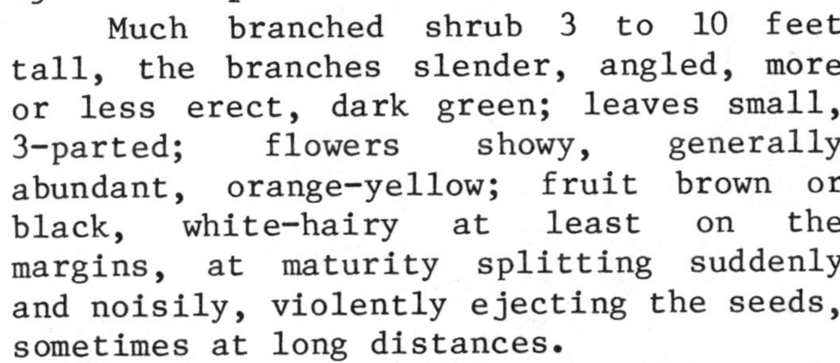

Much branched shrub 3 to 10 feet tall, the branches slender, angled, more or less erect, dark green; leaves small, 3-parted; flowers showy, generally abundant, orange-yellow; fruit brown or black, white-hairy at least on the margins, at maturity splitting suddenly and noisily, violently ejecting the seeds, sometimes at long distances.

x 1/2

Scotch Broom was introduced, by early settlers of the Pacific Coast, as a garden ornamental, and since that time has spread far beyond the bounds of cultivation until it now covers many acres of wasteland. Undoubtedly the first introduction was supplemented by others, for it is now scattered over inland valleys as well, and even east of the Cascades. It has also been deliberately planted as a soil binder on highway cuts and fills.

Not only has the shrub become a nuisance in pastures and cultivated fields, but it has crowded out the native vegetation characteristic of the state, often forming pure stands for miles along highways and country roads.

GORSE

Ulex europaeus L.

A stiff much branched shrub, 1 to 9 feet tall, the branches often ending in spines and bearing spine-like leaves; flowers deep orange-yellow, abundant; fruit dark, more or less covered by long white hairs.

This shrub, like Scotch Broom, is a native of Europe and was first brought to the Pacific Coast as a garden shrub. It, too, has escaped from cultivation and has become a pest on range lands, particularly along the central and southern Oregon coast. Both Scotch Broom and Gorse may become serious fire hazards during the dry summer months. When Bandon, Oregon, was largely destroyed by fire in 1936, the acres of pure stands of Gorse offered a ready path from the source of the blaze to the town, and were largely responsible for the rapidity with which the fire swept over the area.

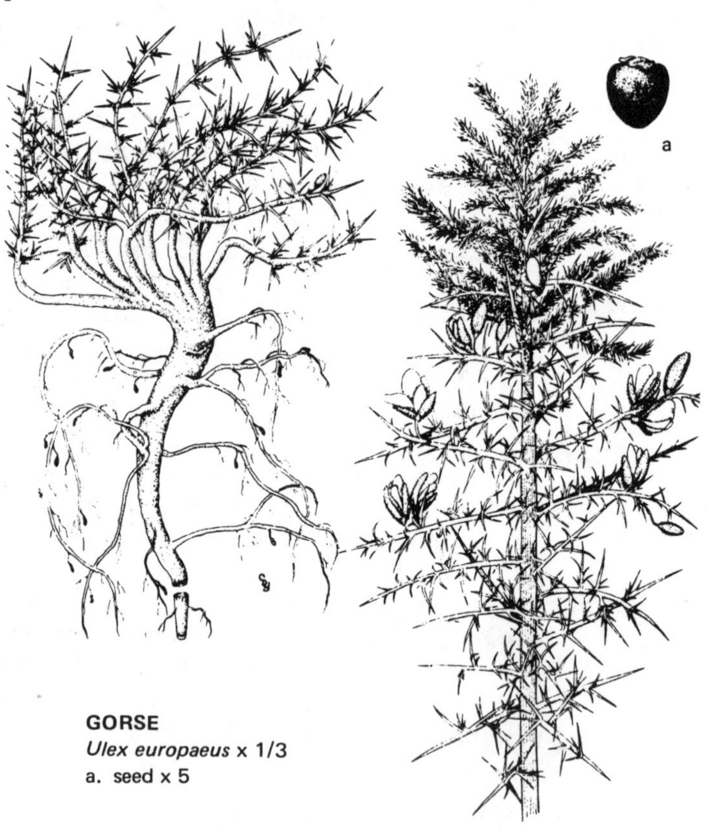

GORSE
Ulex europaeus x 1/3
a. seed x 5

SHOESTRING PSORALEA

Psoralea lanceolata Pursh

Perennial; stem 1 to 2 feet tall, simple or branching, erect or spreading; leaflets in 3's, narrow, minutely black-dotted, with a strong odor; flowers white or bluish, in short-stalked compact spikes borne in the leaf axils; fruit very small, one-seeded.

Shoestring Psoralea, so named from the conspicuously long slender tough underground structures from which new plants arise, is a common inhabitant of the plains and prairies of the Middle West, whence it has spread to similar areas in the Pacific Northwest. In recent years it has come to be a pest in cultivated fields east of the Cascades, its strongly entrenched underground parts making it a formidable rival of most crop plants. The leaflets are shaped much like those of alfalfa, and seeds of these two species are similar in size.

SPOTTED LOCO
Astragalus lentiginosus x 2/3
a. fruits x 2/3

SPOTTED LOCO, LOCOWEED or RATTLEWEED

Astragalus lentiginosus Dougl.

Perennial; stems clustered, erect or spreading, 1/3 to 1 foot long; leaves 1/2 to 3 1/2 inches long, the leaflets 11 to 19, elliptic to oblong, arranged along 2 sides and at the end of the axis; flowers generally 1/2 inch or less long, somewhat clustered on the stalks, white, sometimes purple-tinged; fruit 3/4 to 1 inch long, greatly inflated, curved, somewhat flattened from front to back, gradually narrowed from middle to apex into a slender beak, surface of the fruit smooth or with a few appressed hairs, and often mottled with purplish brown.

Various species of Loco (or Milk Vetch) are common and sometimes abundant on the sheep and cattle ranges of the Pacific Northwest. They vary greatly in appearance but all have compound leaves (divided into leaflets) and pea-shaped flowers borne in clusters of few to many. The fruits may be shaped like those of pea, or be greatly inflated around the seeds, giving rise to the name Rattleweed. In some species a false partition divides the cavity of the pod into two compartments rather than one, as is usual for the family. Certain species are poisonous, while others are grazed with no apparent harmful effects. Spotted Loco is one of the poisonous species.

It is impractical here to describe the many species of *Astragalus*. The two here discussed are taken as representative of the genus; but the degree of variation existing in the fruits of different species is suggested in the accompanying drawings.

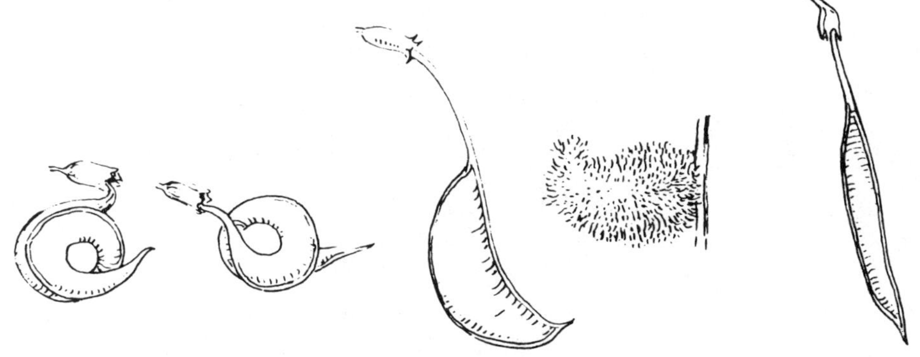

LATE MILK VETCH

Astragalus serotinus Gray

Perennial; stems slender, branching, generally many from a woody base, erect or spreading; leaflets very slender, 11 to 21 per leaf, covered beneath by stiff hairs lying parallel to the surface, smooth above, pointed at the apex; flowering stem slender, extending beyond the leaves, loosely 5- to many-flowered; flowers small, white with purple-tipped keel; fruit slender, 5/8 to 1 inch long, straight, nearly tubular, 2-ridged, spreading outward at maturity.

This is a poisonous species of *Astragalus* which is found on the ranges east of the Cascade Mountains.

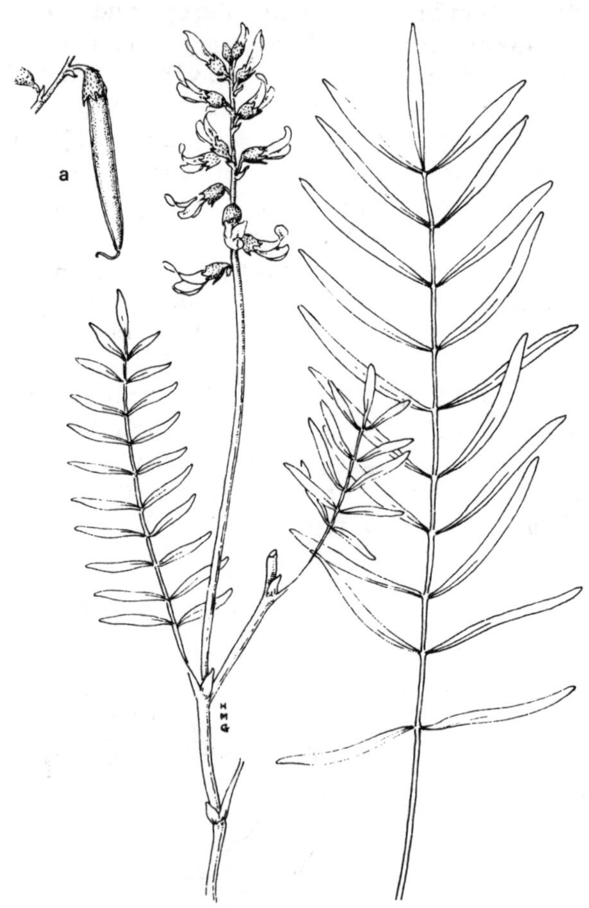

LATE MILK VETCH
Astragalus serotinus x 1/2
a. fruit x 1

EVERLASTING or PERENNIAL PEAVINE or WILD SWEET PEA

Lathyrus latifolius L.

Perennial; stems broadly winged, 2 to 7 feet long, of more or less climbing habit; leaflets 2, broadly lance-shaped, stipules 1 to 2 inches long; tendrils well developed; flowers 5 to 15, approximately 1 inch long, white, pink or red (occasionally striped); fruit 10-25-seeded, 2 to 4 inches long.

A native of Europe, this is our common weedy sweet pea.

FLAT PEAVINE

Lathyrus sylvestris L.

Similar to *Lathyrus latifolius* but with 5 to 9 red flowers about 5/8 of an inch long, and with narrow lance-shaped leaflets; fruits 1 1/2 to 2 1/2 inches long.

This species is also a native of Europe and is frequently established in waste areas, especially along roadsides.

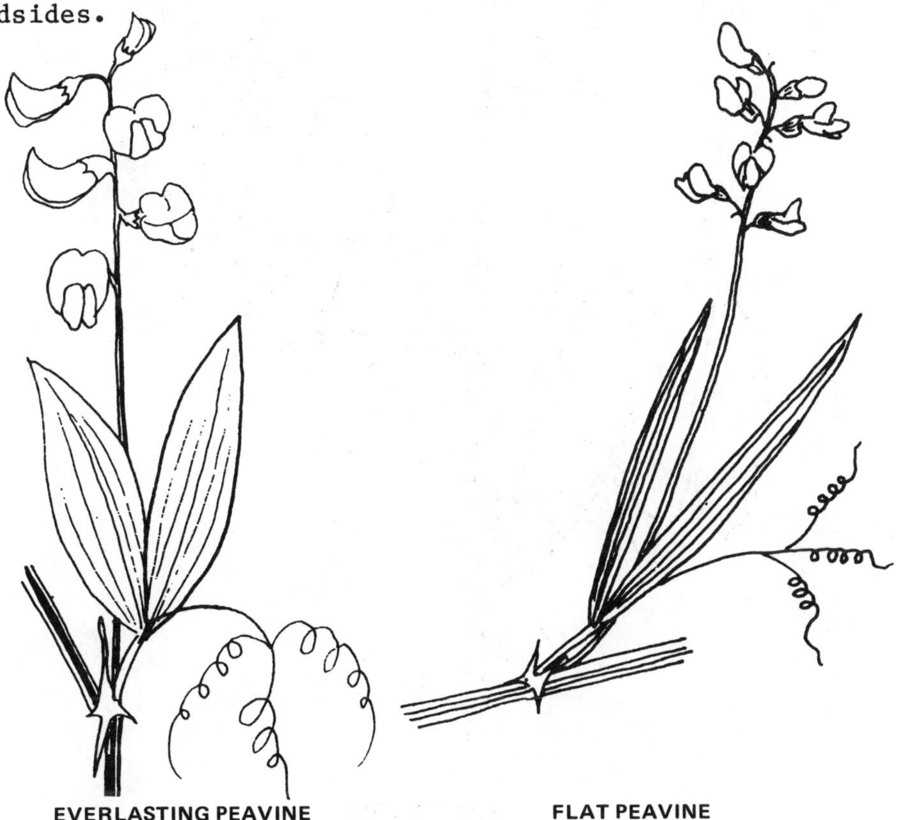

EVERLASTING PEAVINE
Lathyrus latifolius x 1/3

FLAT PEAVINE
Lathyrus sylvestris x 1/3

HAIRY VETCH
Vicia villosa x 1

HAIRY VETCH

Vicia villosa Roth

"Hairy" annual or biennial; stems 1 1/2 to 6 feet tall; leaflets 10 to 20, linear to narrowly lance-shaped, 3/4 to 1 inch long; tendrils well developed; racemes usually 1-sided and 20- to 60-flowered; flowers purplish-red, 3/4 to 1 inch long; fruit 3/4 to 1 1/4 inches long, several-seeded.

Introduced from Europe and commonly escaped and naturalized, especially along roadways.

COMMON VETCH

Vicia sativa L.

Perennial 1 to 2 1/2 feet tall; leaflets 10 to 14, lance-shaped, 3/4 to 1 1/4 inches long; flowers 1 to 3 in the leaf axils, 3/4 to 1 inch long, purplish, often with some red on 2 of the petals; fruit 1 1/8 to 2 3/4 inches long.

This is an extremely variable species; introduced from Europe and now widespread in the United States.

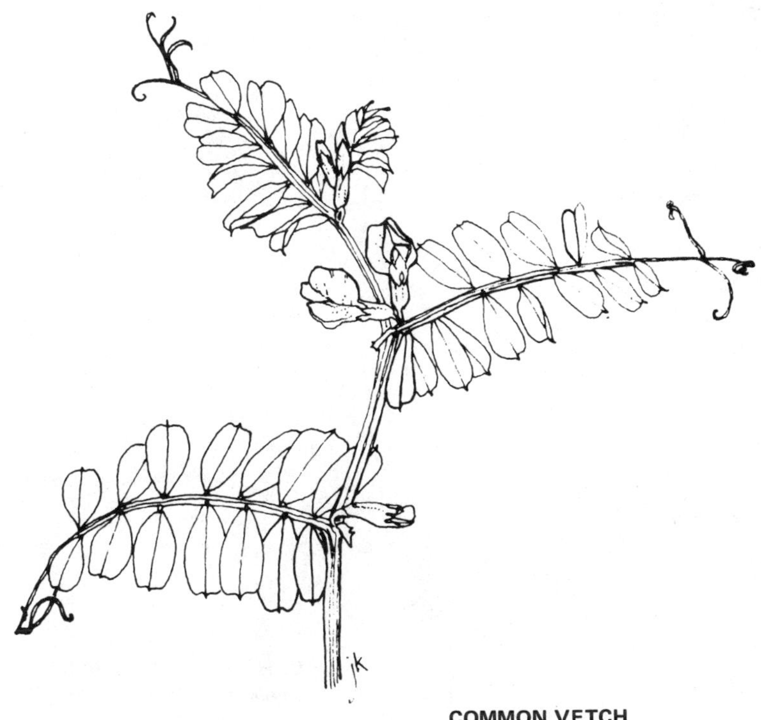

COMMON VETCH
Vicia sativa x 1/2

a. **YELLOW SWEETCLOVER**
Melilotus officinalis x 1

b. **LARGE-LEAVED LUPINE**
Lupinus polyphyllus x 1/2
c. fruit x 2/3

WHITE SWEETCLOVER

Melilotus alba Desr.

Annual or biennial to 9 feet in height; stems much branched; leaves of 3 leaflets, the leaflets toothed; flowers small, white, numerous, usually borne in one-sided racemes; fruit usually 1-seeded; seed dull yellowish or brown.

Eurasian species widespread in the United States as a weed of roadsides and waste places, particularly in gravelly soil.

Melilotus officinalis (L.) Lam. (Yellow Sweetclover) is similar to *Melilotus alba* but the flowers are yellow rather than white. It occurs in similar habitats, but is less abundant than White Sweetclover.

LARGE-LEAVED LUPINE

Lupinus polyphyllus Lindl.

Perennial; stems stout, often reaching 3 to 4 feet; lower leaves long-stalked, forming a large clump; flowers borne in long spike-like racemes, and ranging in color from white through deep blue to dark purple; fruits 1 1/2 to 2 inches long, hairy.

Nearly every portion of the Northwest, including marshes, dry plains, high mountains, fertile valleys, and sandy beaches, has one or more common species of Lupine. In addition to their pea-like blossoms which are borne in racemes, and to their smooth or hairy pea-pod-like fruits, these plants are distinguished by their leaves which are cut into 4 to 20 narrow leaflets radiating spoke-like from the apex of the leaf stalk.

Certain species of Lupine are known to be poisonous to livestock. Not all species are poisonous and those that are often vary seasonally in toxicity.

The single species here described is taken as representative of the genus. Large-leaved Lupine is a well-known species in the Cascades and through western valleys, also in mountains east of the Cascades. This is one of the largest species in the Pacific Northwest. The Small-flowered Lupine (*Lupinus micranthus* Dougl.), a common annual roadside weed, rarely exceeds a foot in height; while certain high altitude and sea beach species spread prostrate on the ground.

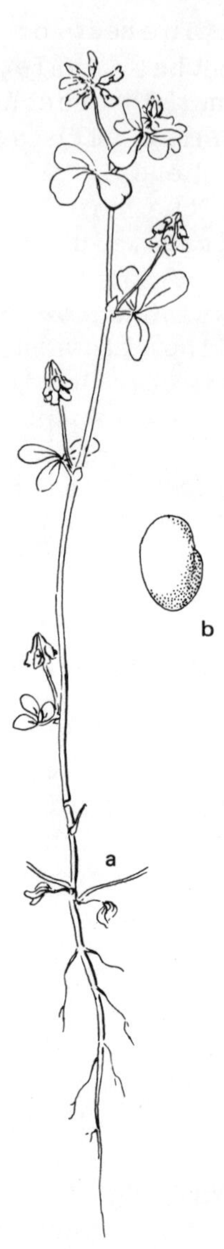

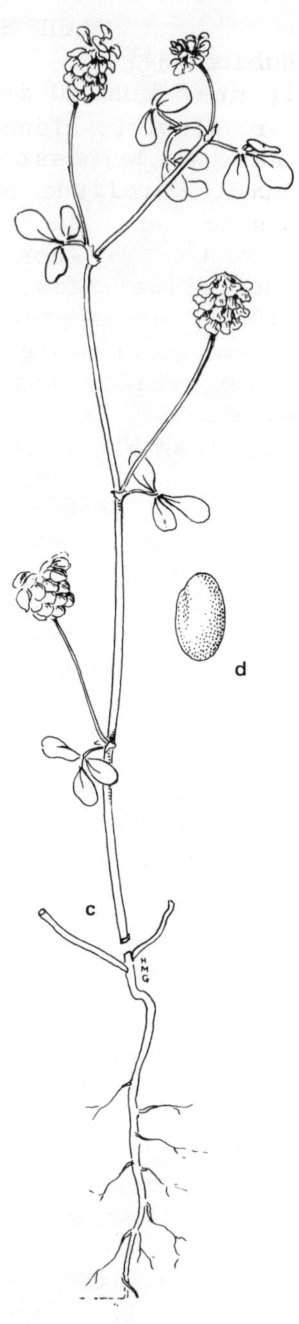

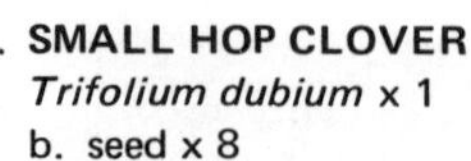

a. **SMALL HOP CLOVER**
Trifolium dubium x 1
b. seed x 8

c. **LOW HOP CLOVER**
Trifolium procumbens x 1
d. seed x 8

SMALL HOP CLOVER

Trifolium dubium Sibth.

Annual; stem 2 to 10 inches long, erect or prostrate, branching or simple, slender, somewhat leafy; leaflets generally narrow, broadest and minutely toothed at the apex, narrowed toward the base, usually with a few hairs on lower side of the midrib; heads yellow, small (generally 3/8 inch or less in width), slightly elongated, the individual flowers few, turning backward with age, not becoming inflated or papery.

This is an irritating weed in lawns, and is common along roadsides and in waste ground. It was introduced from Europe.

Somewhat resembling it, and likewise introduced, is the somewhat rarer LOW HOP CLOVER (*Trifolium procumbens* L.). This is a larger species, the stems sometimes reaching 1 1/2 feet in length or height. In this species, also, the heads are yellow but are larger, often elongating to 1/2 inch; the flowers are more numerous per head and become inflated and papery as the fruits mature. The leaflets are generally broader than in Small Hop Clover, and more conspicuously heart-shaped at the apex.

Both species make yellow unsightly patches in lawns.

BLACK MEDIC

Medicago lupulina L.

x8 jk

Annual; stems 4 to 16 inches long; leaflets 3, broadest at or above the middle; flowers yellow, 10 to 40, in crowded spike-like racemes; fruit 1-seeded, strongly-veined, black at maturity and curved to a kidney shape.

Introduced from Europe and widespread in sandy or gravelly soil.

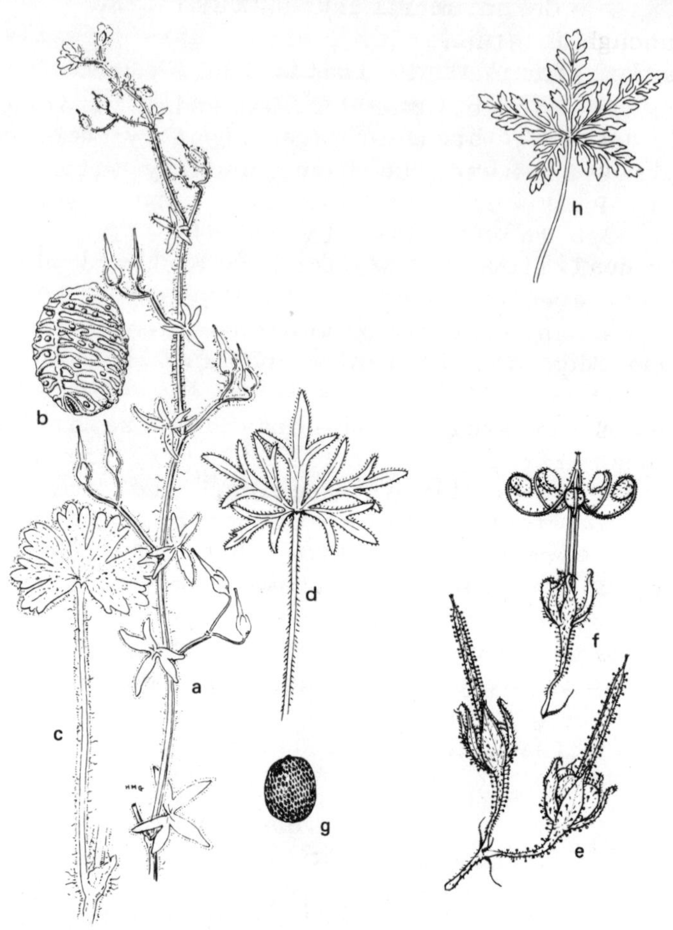

DOVEFOOT GERANIUM
Geranium molle
a. flowering branch x 2/3
b. segment of fruit without beak x 14
c. basal leaf x 2/3

CUTLEAF GERANIUM
Geranium dissectum
d. leaf x 2/3
e. mature fruits x 2
f. fruit separating x 2
g. seed x 2

ROBERT'S GERANIUM
Geranium robertianum
h. leaf x 2/3

Geranium Family: GERANIACEAE

Although familiarity with this family arises principally from the cultivated ornamental plants commonly called Geranium (in reality *Pelargonium*), several native and a few introduced representatives commonly occur in our region.

A conspicuous characteristic of the family is in the fruit which, before maturity, is 5-lobed with a long furrowed beak. As the seeds ripen, the beak elongates further and the entire structure eventually splits into 5 separate long-beaked segments, each containing a single seed. The slender beaks coil and uncoil with changes in humidity, and act as corkscrews which assist in distributing and planting the seeds.

DOVEFOOT GERANIUM

Geranium molle L.

Annual or biennial; stems generally several from the base, weakly erect or more often spreading, more or less covered by fine short hairs; leaves velvety-hairy, 5- to 7-lobed, lobes of the lower leaves deeply toothed, those of the upper leaves few, narrow, and entire; flowers small, reddish purple, paired on stalks borne in the upper leaf axils and exceeding the leaves in length; fruits brown at maturity, the lobes separating, each lobe horizontally wrinkled (as seen by a hand lens), the seed within it smooth-surfaced or nearly so.

This European plant is a common weed west of the Cascades. It develops rapidly in early spring, often from a rosette of leaves already formed during the previous season, and it matures and sows its seeds almost before the ground is ready for tilling. Seeds apparently germinate whenever conditions are favorable, and a succession of plants may be produced throughout the growing season. The common name of this species is derived from the shape of the upper leaves. Several other species of small-flowered Geranium, most of them likewise originating in Europe, occur in the Far West as lawn and garden weeds. They differ principally in leaf characters and in surface of the fruits. Cutleaf Geranium (*Geranium dissectum* L.) is a well-known example.

A rare introduced species, which in a few localities has taken over cultivated fields, is Robert's Geranium

(*G. robertianum* L.). It is distinguished by small showy bright reddish flowers and greatly divided leaves, as illustrated. A native of Europe, it probably was introduced as a garden plant, but having escaped from cultivation it has become widely established in a few limited areas.

REDSTEM FILAREE

Erodium cicutarium (L.) L'Her.

Annual or biennial; stems 1 inch to 2 feet in length, spreading or erect, generally from a rosette of basal leaves; leaves divided feather-like into narrow lobed or toothed segments, both leaves and stems hairy; flowers purplish-pink, generally borne in clusters of 2 to many (in very small plants, sometimes borne singly); fruit 5-lobed, long-beaked, each lobe finally splitting away and carrying with it a long section of beak.

A native of Europe or adjoining Asia, Redstem Filaree is now known on every continent. It occurs over much of the Pacific Northwest, and is one of the earliest plants to blossom in the spring. On railroad beds, hard-packed roadsides, or in open fields where the sun strikes warmly, tiny flattened rosettes of leaves may appear, each sending up a few flower stalks no more than an inch in height. Their small size may be offset by their abundance, and entire hillsides are painted rose-color by them. Under favorable soil and weather conditions, Redstem Filaree may reach a height of two feet or more, with leaf size in proportion.

The beaks of individual units of the fruits, after the five split apart, coil and uncoil with changes in the moisture content of the air helping to thrust the seeds into the ground.

This species, together with several others introduced into western America, has sometimes been grown for forage. They are noxious weeds only when they crowd out more valuable crops. They may be somewhat troublesome in early gardens.

REDSTEM FILAREE
Erodium cicutarium x 1 1/3
a. beaked segment of fruit x 2

WOODSORREL
Oxalis stricta x 1

Woodsorrel Family: OXALIDACEAE

Herbs (in our area) with watery sour juice; flowers delicate, sepals and petals 5, stamens 10, styles 5; leaves alternate and palmately compound with 3 heart-shaped leaflets (in ours).

CREEPING WOODSORREL

Oxalis corniculatus L.

Taprooted perennial; stems trailing and usually freely rooting at the nodes; leaflets 1/4 to 1 inch long, green, bronze, reddish or purple; flower stalks 1- to 5-flowered; petals yellow, 1/8 to 3/8 inch long; fruit hairy, 3/8 to 1 inch long, splitting at maturity to release the brown, wrinkled seeds.

This European native is widespread as a weed in North America; in our area it is found in waste ground and gardens and is often a problem in greenhouses. Like Mustard and Chickweed, this species can be found in bloom throughout the year.

A similar species, *Oxalis dillenii* Jacq. (Southern Yellow Woodsorrel), is occasional in our area, chiefly east of the Cascades. It differs from *Oxalis corniculatus* in its more erect habit, greener color and its seeds with white ridges.

WOODSORREL

Oxalis stricta L.

Perennial from slender fleshy rootstocks; stems prostrate or more often erect; leaflets gray-green, 3/8 to 3/4 inch broad and up to 2 inches long; flowering stalks 1- to 7-flowered; petals yellow, 1/8 to 3/8 inch long; fruit hairy, up to 1 inch in length; seeds brown, wrinkled.

This native species is found throughout the United States, but is only occasional with us, chiefly west of the Cascades. It occurs in dry open areas and can become troublesome in lawns and gardens.

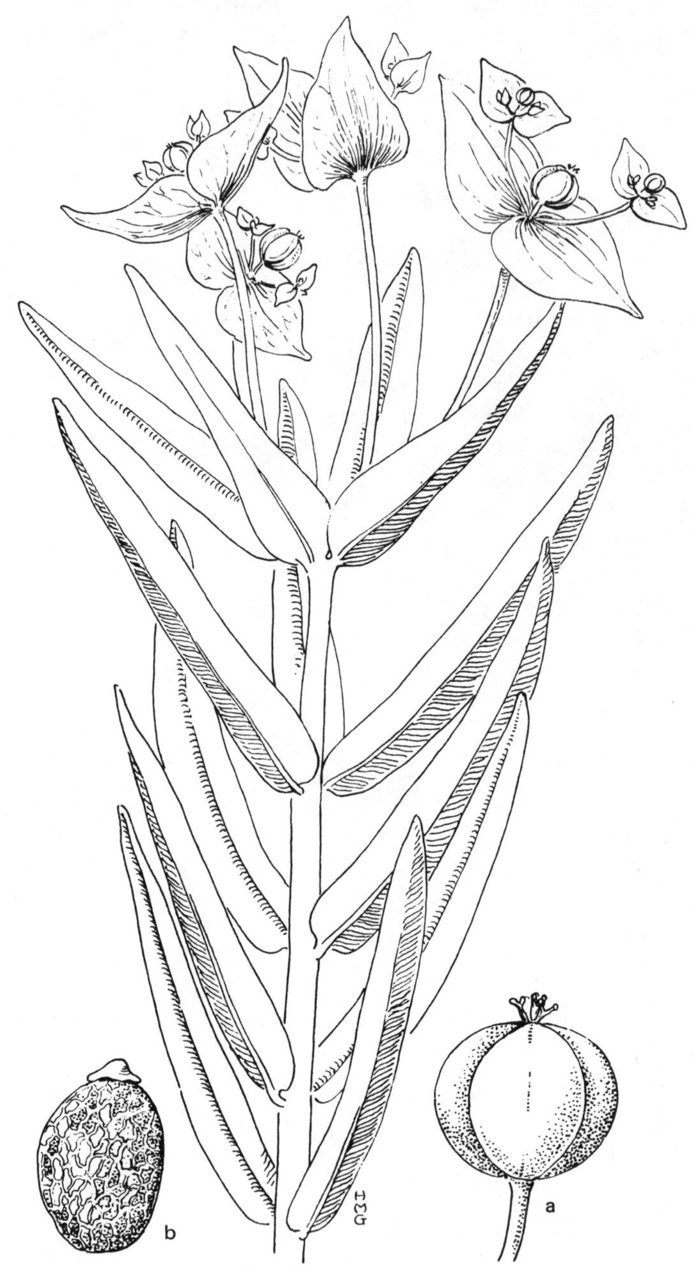

MOLE PLANT
Euphorbia lathyris x 2/3
a. fruit x 3
b. seed x 5

Spurge Family: EUPHORBIACEAE

The Spurge family, from which are derived Para rubber, castor oil, and tapioca, and which supplies the Poinsettia, is represented within our limits by a number of weedy species which, in probably all cases, are poisonous. Most of our species, in common with many other representatives of the family, contain a milky juice.

MOLE PLANT or CAPER SPURGE

Euphorbia lathyris L.

Annual; stem stout, 1 to 3 1/2 feet tall, whitish green, erect, branching above; lower leaves stalkless, long, narrow, arranged in crowded or somewhat distant alternating pairs, or those near the flowering branches sometimes in 4's; upper leaves broad at the base, pointed at the apex; fruit 3-lobed, reaching nearly 2 inches in diameter, seed 5/16 inch long, brownish, rounded at one end, with a conical projection at the other.

Introduced from Europe, perhaps as a cultivated plant, this species has become somewhat naturalized in the Pacific Northwest. It is unimportant, generally, as a weed, but is poisonous both to stock and human beings. When broken, the plant exudes a milky juice, and skin reactions similar to those from Poison Oak may follow its handling by persons sensitive to it, while serious if not fatal digestive disturbances result from eating any part of the plant. Children are attracted by the large fleshy fruits and severe poisoning from this source has been reported.

The plant has been used in gardens for its reputed value in discouraging moles and gophers. Considerable difference of opinion exists regarding its efficacy for this purpose.

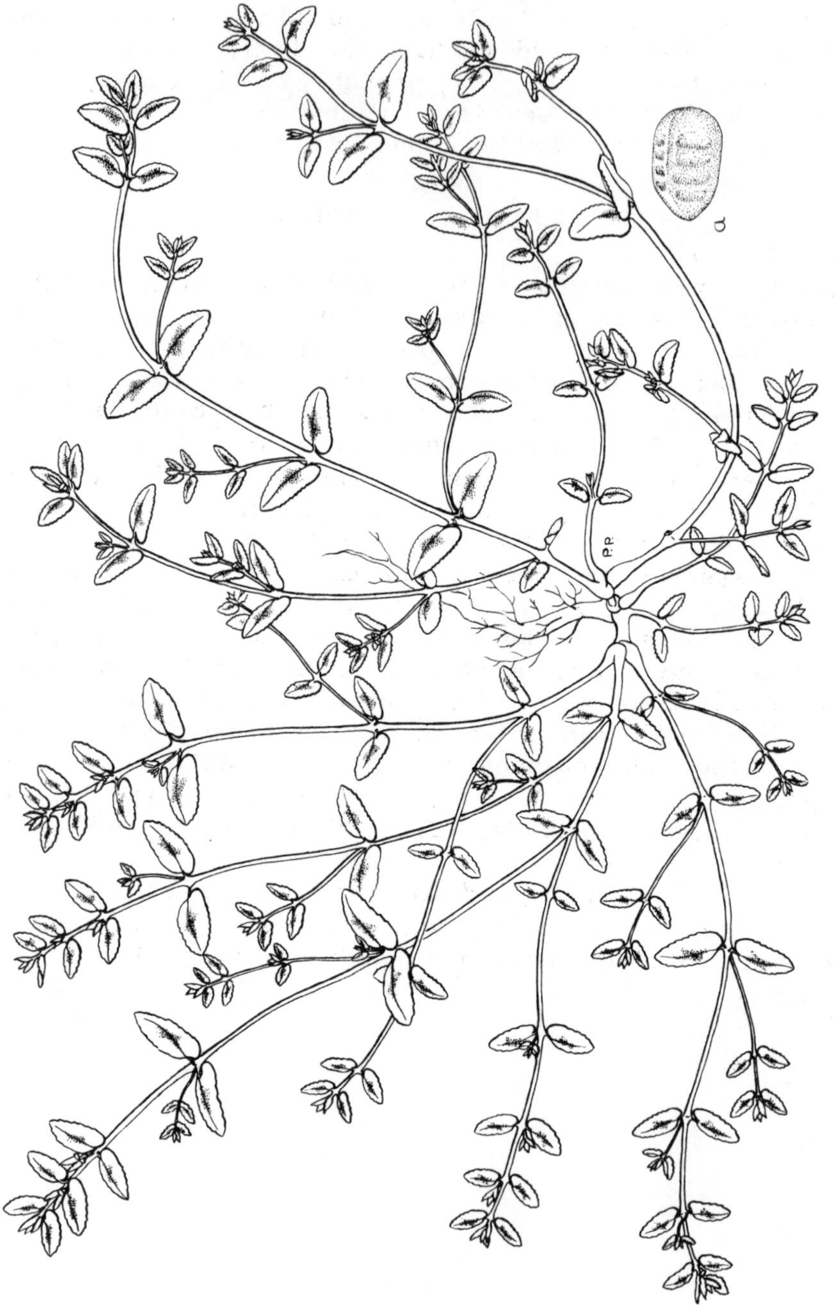

PROSTRATE SPURGE
Euphorbia supina × 1
a. seed × 20

PROSTRATE SPURGE

Euphorbia supina Raf.

Annual; stems hairy, branching from near the base, the branches prostrate or their tips ascending; leaves borne in pairs, 1/4 to 3/4 inch long, about 1/3 as broad, asymmetrical at the base, the margins somewhat finely and remotely toothed, hairy on the under surface and smooth above, generally with a central wine-colored spot; flowers inconspicuous; fruit hairy, 3-angled; seed about 1/25 of an inch long, miscroscopically ridged.

A native of the eastern United States, this species has become sparingly naturalized in the Pacific Northwest and occasionally occurs as a garden weed. Once it has become established, its ability to seed readily and freely makes it difficult to control.

Three other prostrate species of *Euphorbia* are occasional in our area: *Euphorbia glyptosperma* Engelm. (Ridgeseed Spurge) and *Euphorbia serpyllifolia* Pers. (Thyme-leaf Spurge) occur mainly east of the Cascades and differ from *Euphorbia supina* in their smooth stems and fruits. *Euphorbia glyptosperma* has ridged seeds and thick-margined leaves while *Euphorbia serpyllifolia* does not. *Euphorbia maculata* L. (Spotted Spurge) is more common in the eastern United States. It has hairy stems, as does *Euphorbia supina*. Its stems are more ascending, its leaves longer and its fruits are not hairy.

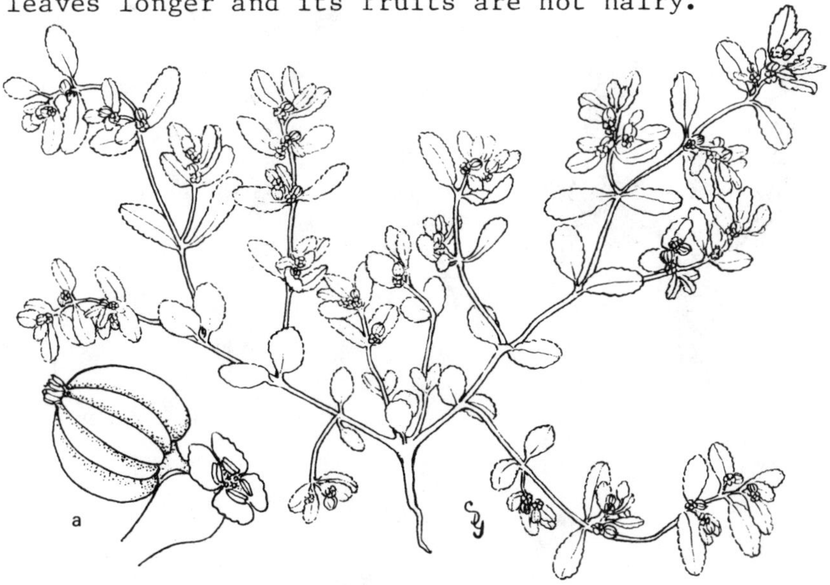

THYME-LEAF SPURGE *Euphorbia serpyllifolia* x 1 a. fruit x 12

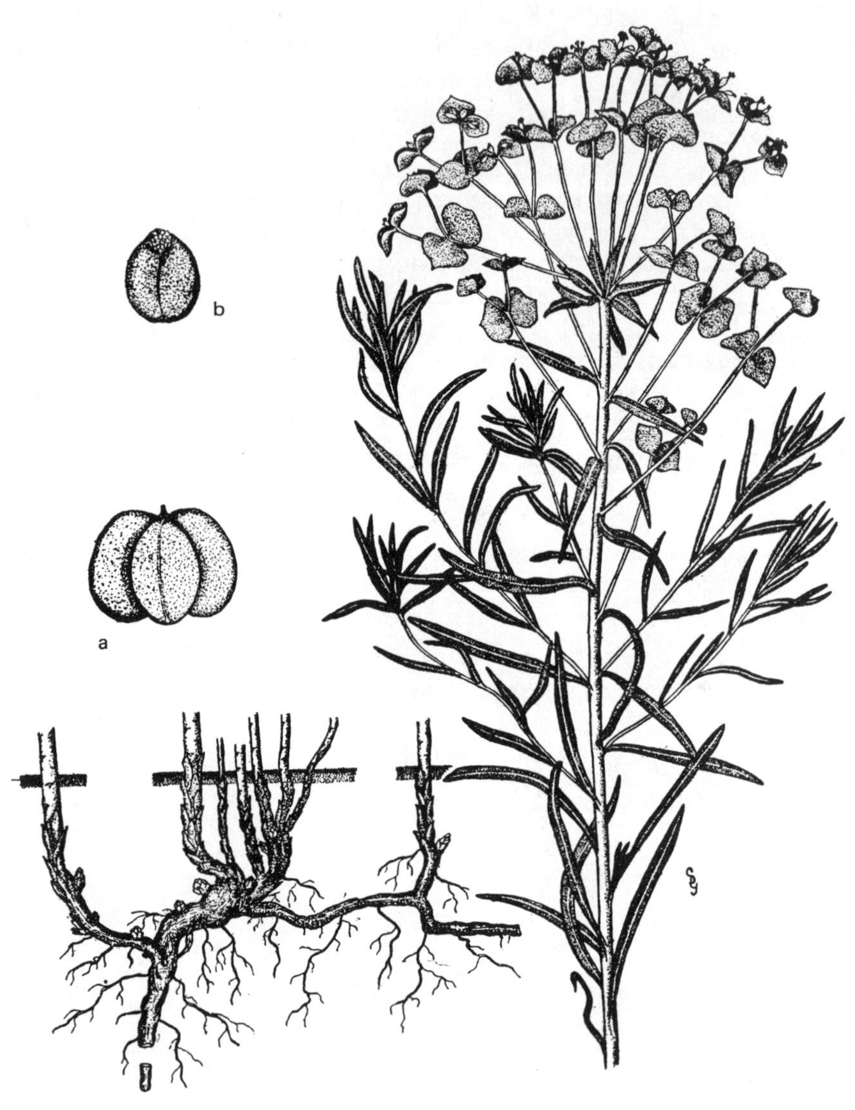

LEAFY SPURGE
Euphorbia esula x 1/3
a. fruit x 6
b. seed x 9

LEAFY SPURGE

Euphorbia esula L.

Perennial; spreading by woody rootstocks and forming colonies; stem erect, generally with numerous weak sterile branches, main stem 3/4 to 3 feet tall, smooth, with a milky juice when bruised; leaves alternate, stalkless, 1 to 4 inches long, narrow, somewhat bluish green; leaves of the flowering branches paired, broad, each pair closely clasping the stem; flowers yellowish, minute, borne on forking leafy branches, the branches and paired leaves becoming yellow at maturity; entire plant turning bright red or orange in the fall; pod 3-lobed with 3 compartments, each lobe opening explosively to eject its single seed; seed oblong, grayish, brownish, or purplish.

This persistent perennial, propagating both underground and by seed, is a serious pest in the Middle West, and has been introduced into the Pacific Northwest, where its possibilities as a menace are sufficiently recognized that the weed laws of Oregon, at least, declare unsalable any small commercial seed containing seed of this plant. Thus far its range in our area is limited, but its presence and spread should be discouraged before it assumes greater proportions.

PETTY SPURGE

Euphorbia peplus L.

Annual; stem simple or branching, 1/3 to 1 foot tall, smooth, exuding a milky juice when bruised; leaves bright green above, paler beneath, lower leaves slender-stalked, blades broadly elliptic, upper leaves on flowering branches stalkless, broadest at the base; fruit 3-lobed, each lobe 2-ridged on the back; seed whitish with several rows of dark pits.

The lower leaves are borne singly at the nodes, some with branches in their axils. Later, 2 or 3 leaves appear at one node, a branch arising from the axil of each, and from this point, paired leaves with paired branches continue above. These are flowering branches, producing inconspicuous flowers and fruit. Seed are produced freely and apparently are capable of immediate germination, for in an irrigated or sprinkled garden, these plants, like Chickweed, are in evidence throughout the growing season. Petty Spurge is reported poisonous both to stock and man.

A European plant, rather commonly found as a weed in gardens and about shrubbery. Seeds in the soil germinate early in the spring, and these plants together with common Chickweed and other quick-growing annuals may form a dense ground cover before the soil can be worked.

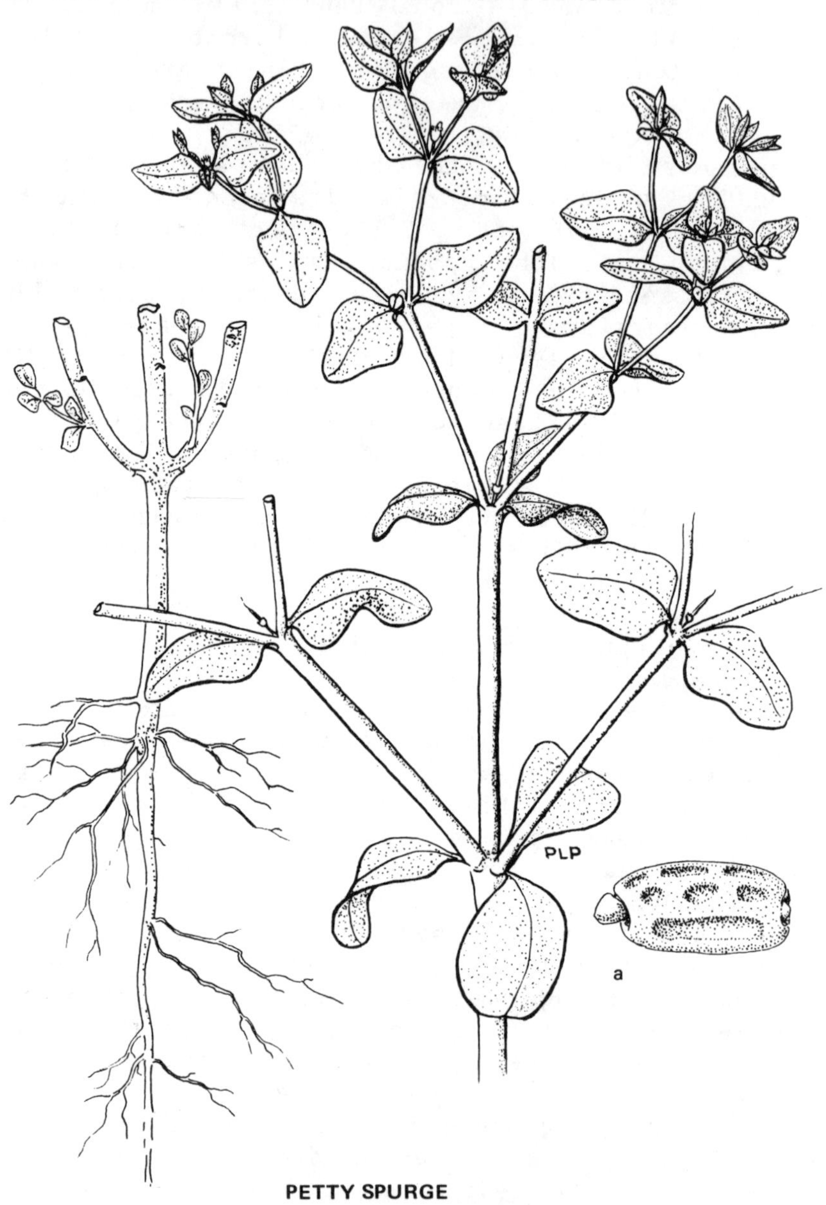

PETTY SPURGE
Euphorbia peplus x 1 1/3
a. seed x 15

TURKEY MULLEIN

Eremocarpus setigerus Benth.

Annual; whole plant grayish green, spreading or prostrate, branching from the base, a single plant covering an area from 3 inches to 2 feet in diameter; leaves thick, broadly ovate to rounded, with 3 main veins from the base, the surface covered by a thick gray layer of minute star-shaped hairs; flowers inconspicuous, of 2 kinds, borne in the lower leaf axils and at the ends of branches; seed one to each pistillate flower, approximately 1/8 inch long, narrow, brown, uniform in color, or mottled.

Turkey Mullein is a native of the Pacific Coast, thriving on dry sandy soil. From sand and gravel bars of rivers it may be introduced, through sandy loam, to gardens where it occasionally gives trouble. Summer fallow areas may become infested by it, though only occasionally does it become a seriously noxious weed.

The plant contains a poison which was utilized by the Indians to stupefy fish in streams, thus facilitating the catch. The minute hairy covering of the stems and leaves is as irritating to many persons as Poison Oak, producing painful redness and swelling. It can well be, however, that this action is at least largely mechanical, since the hairs are easily loosened and readily puncture the skin.

TURKEY MULLEIN *Eremocarpus setigerus* x 1/2 a. seed x 3

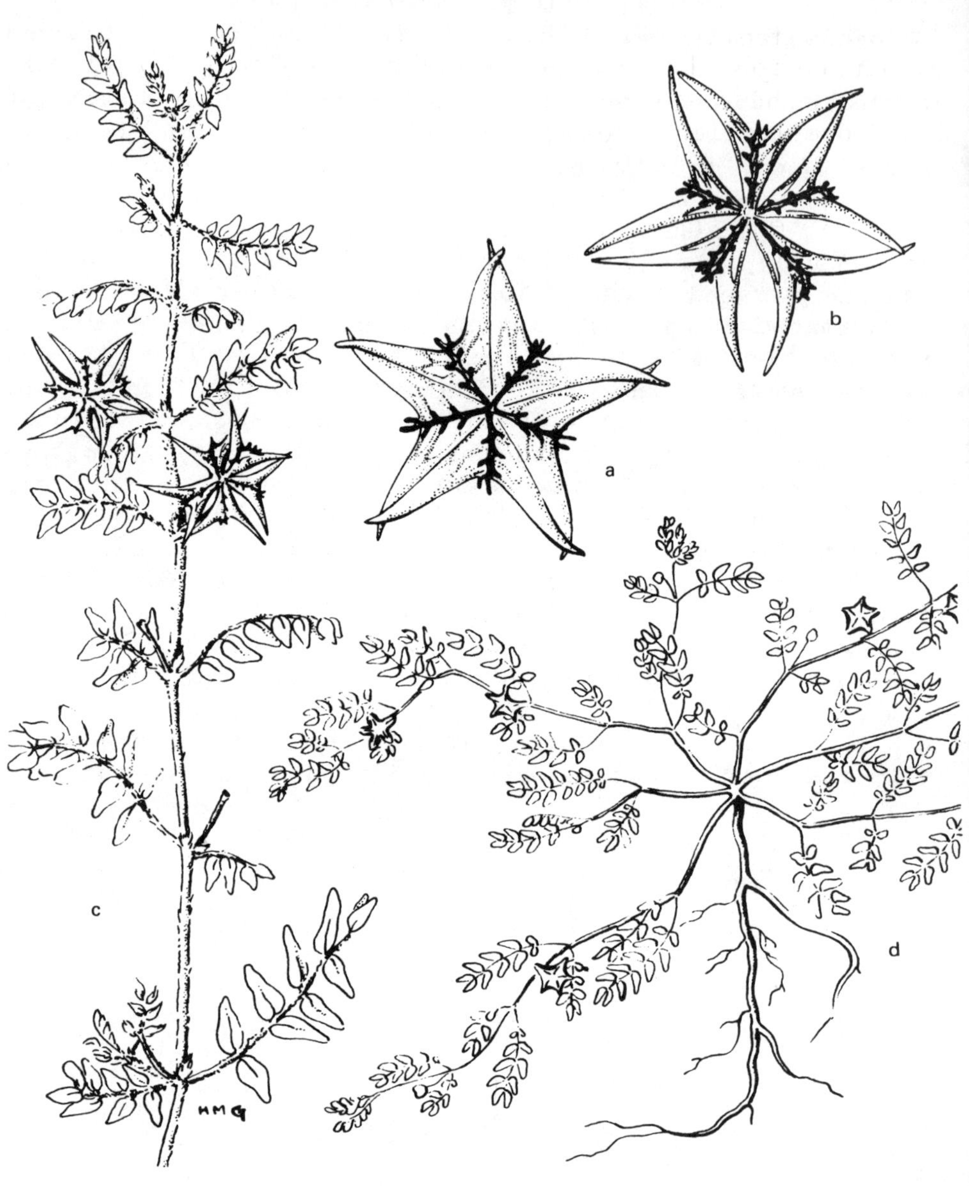

PUNCTUREVINE

Tribulus terrestris

a. fruit before separation x 1 1/3

b. fruit after separation into 5 segments x 1 1/3

c. branch x 2/3

d. habit of plant

Caltrop Family: ZYGOPHYLLACEAE

This family, which has only two known representatives in our region, belongs primarily to the tropics. In most of its members the leaves are paired and each consists of 2 or more leaflets.

PUNCTUREVINE

Tribulus terrestris L.

Annual; stems 6 to 20 inches long, branching from the base, prostrate or somewhat ascending, or even erect when competing with taller plants; leaves divided into 5 to 8 pairs of leaflets, these generally asymmetrical at the base; flowers small, yellow, short-stalked, borne in the leaf axils; fruit consisting of 5 sections, these separating at maturity, each with 2 long woody spines separated by a narrow line of very short projections.

This native of southern Europe has become a nuisance in the Pacific Northwest where it was introduced largely by means of highways and railroads. The spines of the fruit become attached to objects passing over the plant, which thus serve as a means of transportation to other areas. The spines readily penetrate the outer rubber of tires, rendering cars and trucks efficient seed carriers. Fur and hair of animals likewise are so utilized, and the value of wool may be decreased by the quantity of burs present. The spines are injurious to animals and to the bare feet of children.

Sumac Family: ANACARDIACEAE

The Sumac Family contains Poison Oak, Poison Ivy, Sumac, Pistachio, Cashew, and an Asiatic species which is the source of Japanese lacquer. Most members of the family contain a milky or a resinous juice, and several are poisonous.

Wide variation in appearance exists in the family; but in general the individual flowers are small, though sometimes occuring in showy clusters. In the shrubs designated as Poison Oak or Poison Ivy, the leaves are divided into 3 (rarely 5) leaflets.

PACIFIC POISON OAK

Rhus diversiloba Torr. & Gray
(=*Toxicodendron diversilobum* (Torr. & Gray) Greene)

A shrub or small tree, 2 to 12 feet tall, or a woody vine climbing trunks of trees to a height of sometimes 20 to 50 feet; stem grayish or brownish, minutely ridged; leaves with 3 (rarely 5 leaflets), these variable, coarsely lobed or toothed or sometimes with the margin nearly entire, the apex generally rounded but sometimes narrowed; flowers small, greenish, borne in clusters in the leaf axils; fruit shining yellowish or white, longitudinally ridged, borne in clusters.

The two extremes in growth habit, found in this plant, are so distinctive that neither its shrubby nor its vining form suggests close relationship to the other. As a shrub or small tree it may be densely branched and compact, and this fact, together with a fancied resemblance of its leaflets to the leaves of oak, gives it the name of Poison Oak. When, however, it sends long nearly branchless stems far up the trunk of a tree, clinging closely to it by means of aerial roots, the plant is sometimes designated as Poison Ivy. Since it belong to the Sumac Family, it is in reality, of course, neither an oak nor an ivy.

Not only is Poison Oak or Ivy a menace on account of its toxic qualities; it may become a weed pest as well, thriving along roadsides and spreading vegetatively as well as by seeds, over hillsides and fields. It is an attractive member of the native vegetation, particularly in autumn when its foliage assumes brilliant tones of red; but it is one native which we can well do without.

This species occurs principally west of the Cascades, thinning out and generally disappearing toward the coast and at higher altitudes.

The more common species east of the Cascades, *Rhus radicans* L. (Poison Ivy) is more often shrubby than vine-like. Its lateral leaflets differ from those of *Rhus diversiloba* in being stalked, the margins generally entire rather than lobed, and the apex frequently abruptly pointed.

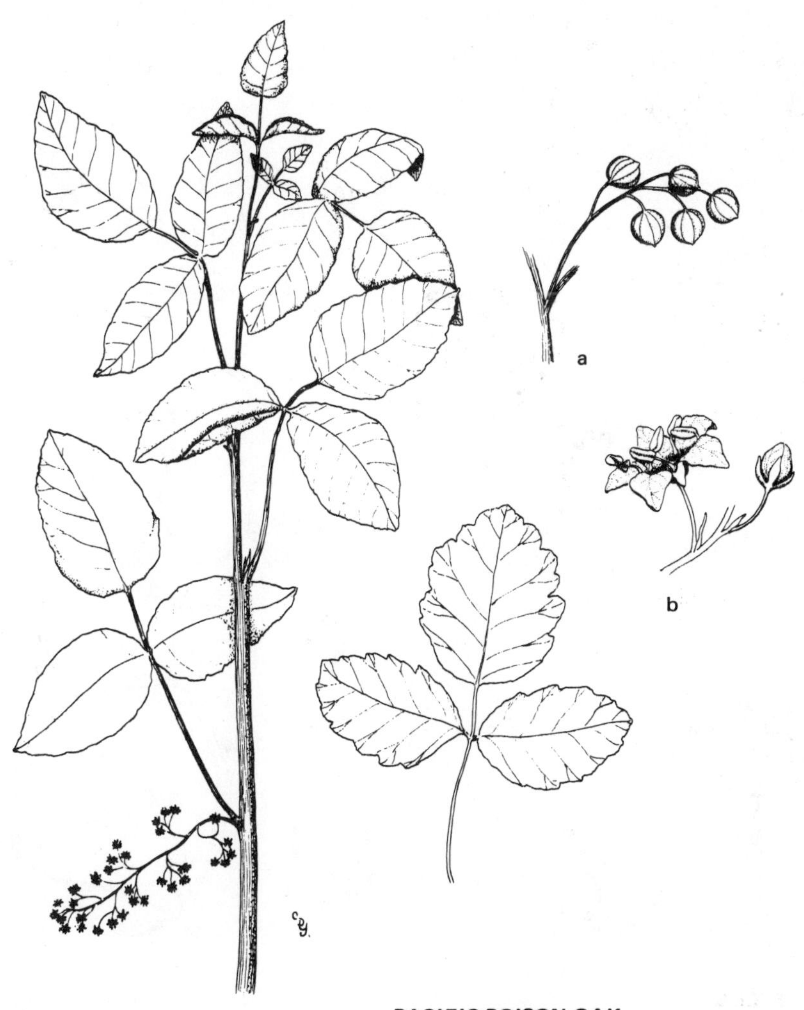

PACIFIC POISON OAK
Rhus diversiloba x 1/2
a. fruits x 1
b. flower and bud x 6

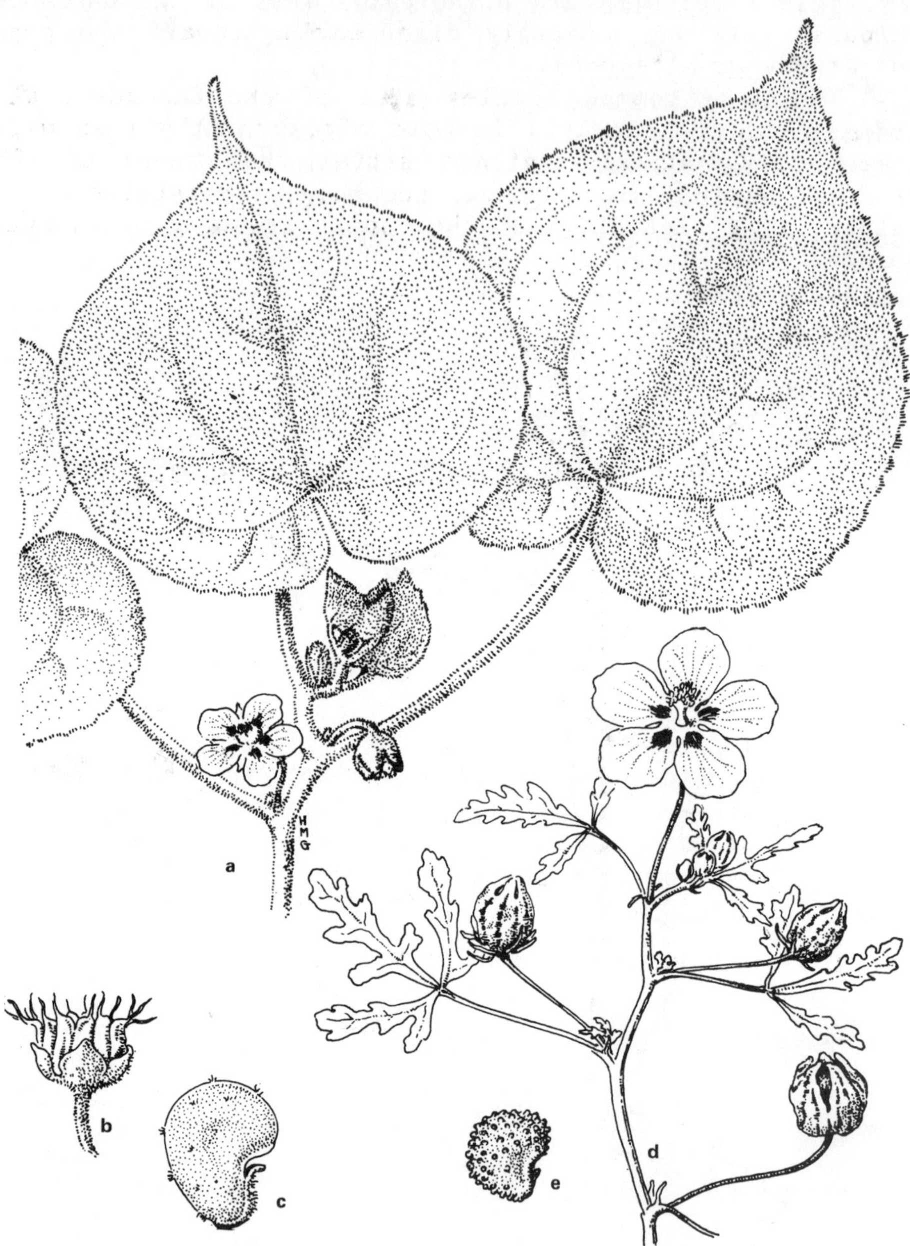

VELVETLEAF
Abutilon theophrasti
a. flowering branch x 2/3
b. cluster of fruits from a single flower x 2/3
c. seed x 6

VENICE MALLOW
Hibiscus trionum
d. flowering branch x 2/3
e. seed x 6

Mallow Family: MALVACEAE

The mallow family has long been known by Hollyhock, Okra, Cotton, Hibiscus, and other ornamental and useful plants. Various native species occur in the Northwest, some of them showy. Several are common along roadsides and in uncultivated fields, but are generally of no significance as weeds. Certain introduced species, however, are troublesome in gardens, orchards, and cultivated field crops.

The flower of Hollyhock is characteristic of those commonly found in the family, with veiny tissue-like petals, and with stamens united in a tube around the pistils.

VELVETLEAF

Abutilon theophrasti Medic.

Annual; 2 to 4 feet tall; stem stout, erect; leaves broadly heart-shaped, coarsely to minutely toothed, velvety-hairy, 2 to 8 inches wide, abruptly pointed at the apex; flowers yellow, 1/2 to 3/4 inch in diameter; fruit 1/2 to 3/4 inch high, 3/4 to 1 inch wide, consisting of 10 to 15 units which separate at maturity, each with a style and each splitting at maturity to release the seeds; seed 1/8 inch or more in diameter, purplish black, curved, minutely warty.

Naturalized from southern Asia and as yet only sparingly found in our area, it is sometimes reported as a weed of cultivated ground. The longevity of its seeds which are borne in abundance, makes it a potential menace.

VENICE MALLOW, FLOWER-OF-AN-HOUR or BLADDER KETMIA

Hibiscus trionum L.

Annual; plant more or less stiff-hairy; stems 1/2 to 1 1/2 feet long, erect or more often spreading, generally branching from the base; leaves deeply cut into 3 or 5 lobes, the middle lobe often considerably longer than the laterals (all coarsely toothed); flower reaching 1 1/4 inch in diamater, yellow with a purple or blackish center, the petals falling quickly leaving the sac-like fruit enclosed within the calyx, this enlarging as the fruit develops, becoming papery and bladdery, conspicuously veined; seed about 1/12 inch long, dark

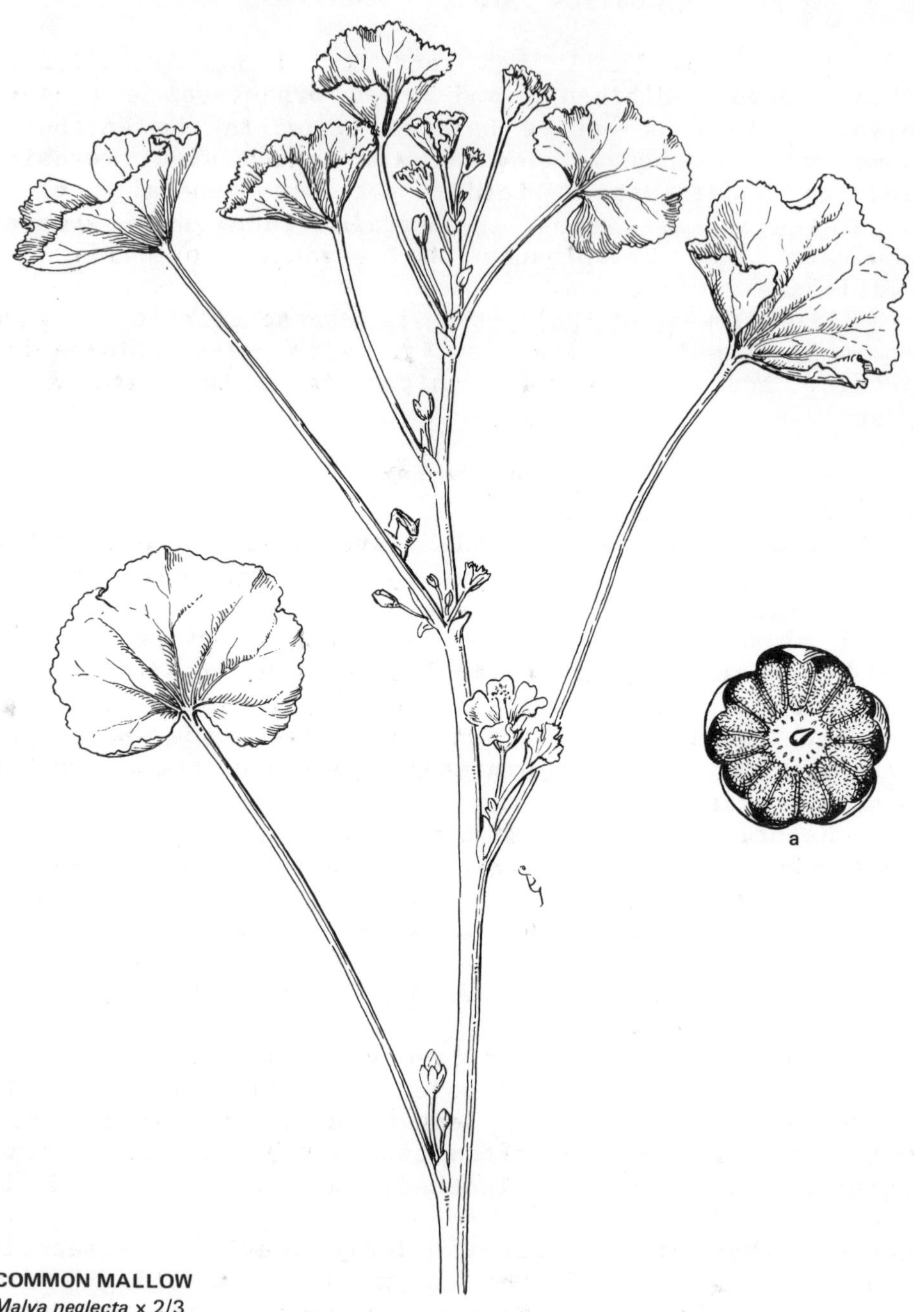

COMMON MALLOW
Malva neglecta x 2/3
a. fruits from a single flower x 4

brown, slightly curved, minutely warty.

A European plant which has long been known in waste ground of the Eastern States, this species is occasionally reported in our area. In orchards and cultivated fields it sometimes reaches pest proportions. Details of the flower clearly indicate the relationship of this plant to ornamental species and varieties of *Hibiscus*.

COMMON MALLOW or CHEESEPLANT

Malva neglecta Wallr.

Annual, winter annual or biennial; stems generally low-spreading, the tips of the branches erect, 2 to 30 inches long, leafy; leaves long-stalked, rounded with a heart-shaped base, 3/4 to 3 1/2 inches in diameter, inconspicuously 5- to 7-lobed, the lobes with rounded teeth; flowers stalked, white to pale lavender, the petals at least twice the length of the calyx; fruit consisting of a circle of generally one-seeded rounded lobes fused at first but separating at maturity.

This species is a European weed long known in this country, and common in waste places, gardens and cultivated ground. In our region it occurs on both sides of the Cascades.

A similar European species, *Malva parviflora* L. (Little Mallow), is less common in our area and differs from *Malva neglecta* in that its petals are shorter to only slightly longer than the calyx, and with the lobes of the fruit flattened and wrinkled.

St. JOHNSWORT
Hypericum perforatum
a. seed x 15

St. Johnswort Family: HYPERICACEAE

Both native and introduced members of this family occur in the Pacific Northwest but only the species here described is of concern, generally, as a troublesome weed.

In all our species, the flowers are yellow, with many stamens, these typically organized into 3 groups within the flower.

A cultivated species, *Hypericum calycinum*, is an attractive sub-shrub commonly used as a ground cover for banks or borders. Its shining foliage and large golden flowers with a mass of yellow stamens give it a deserved popularity. But this species, too, spreads rapidly by means of stolons or rootstocks and, when it gets out of bounds, may become a nuisance.

ST. JOHNSWORT, KLAMATH WEED or GOAT WEED

Hypericum perforatum L.

Perennial; stem often reddish, woody at the base, propagating vegetatively by underground or surface "runners;" main stems erect, single or several from the base of the plant, accompanied by many slender weak sterile basal stems; leaves numerous, small, narrow, stalkless, borne in alternating pairs the full length of the stem, each leaf when held to the light appearing as if full of pin pricks; flowers 3/4 inch in diameter, bright yellow, borne in branched clusters at the summit of the stems; petals 5, often edged with minute black dots; stamens many, arranged in 3 groups; fruits at maturity splitting into 3 segments to allow escape of the numerous seed; seed less than 1/20 inch long, narrow, microscopically pitted.

St. Johnswort is a native of Europe. Since its introduction many years ago it has become one of our most pernicious pasture weeds. It is particularly abundant west of the Cascades and in certain sections east of the mountains where it is rapidly spreading to previously uninfested areas.

The plant is poisonous to stock but is not eaten when grass is present in sufficient quantity. This situation, however, may result in over-grazing of desirable pasture plants and consequent increased invasion by St. Johnswort. The weed is not a serious threat in cultivated fields.

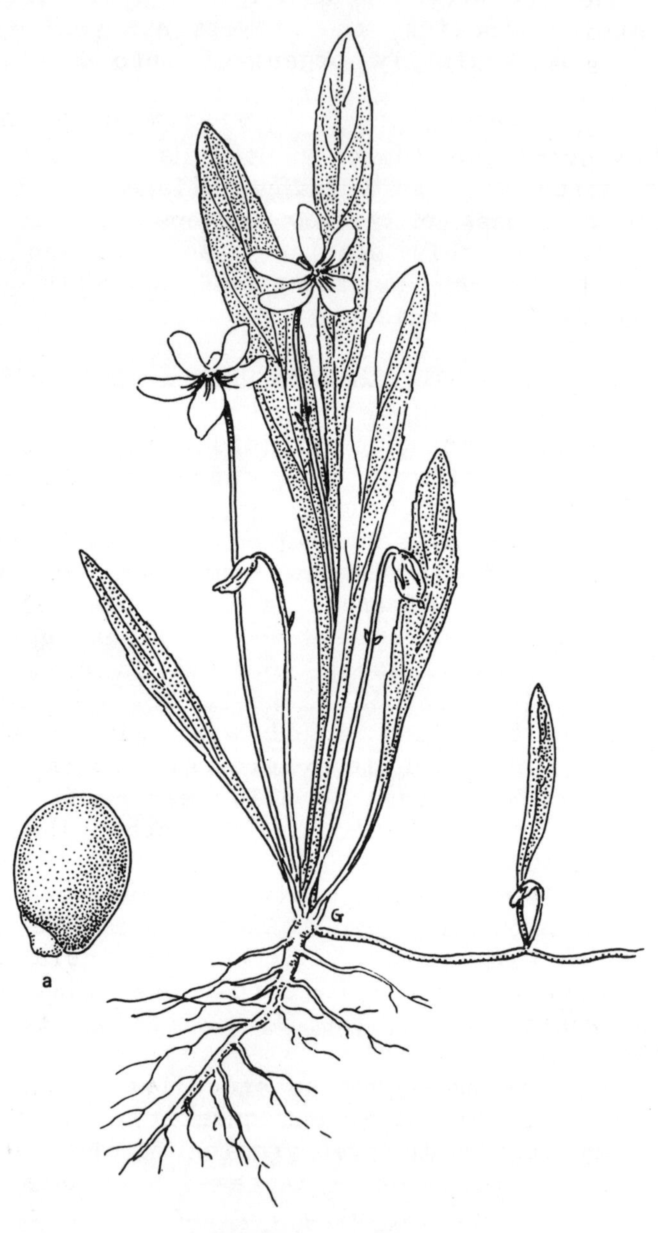

WATER VIOLET
Viola lanceolata x 4/5
a. seed x 12

Violet Family: VIOLACEAE

A well-known family of which many native species occur in our area. The native violets cause no trouble, except that certain of them may become overambitious when introduced into a flower border, but the eastern species here described has become a pest in western cranberry bogs.

The flower in the genus *Viola*, to which violets and pansies belong, is characterized by three shapes and sizes of petals--2 upper, 2 lateral, and a larger basal one.

WATER VIOLET

Viola lanceolata L.

Perennial by stolons; leaves smooth, 1 to 4 inches long, narrow, pointed at the apex, narrowed at the base into a long stalk; flower stalks generally as long as or longer than the leaves; flowers white, the lower and lateral petals purple-veined; fruits produced both by regular flowers and by inconspicuous flowers hidden at the base of the plant.

Water Violet is a native of the Eastern and Middle Western States, but apparently in shipments of cranberry stock it has found its way to commercial cranberry bogs of the Far West. Its ability to creep by stolons, sending up new plants as it goes, permits it to become thoroughly entrenched in a bog within a short time, and control means are difficult to apply in the specialized situation in which it grows.

A few garden varieties of violet may occasionally become troublesome by spreading into borders and lawns. Most species seed freely, and their habit of "shooting" the seed some distance by violent bursting of the fruit enables them to travel rather rapidly over a wide territory.

FIREWEED
Epilobium angustifolium x 1

Evening-primrose Family: ONAGRACEAE

This family is perhaps best known by Fuchsia. Members of the family usually have 2 or 4 petals and an inferior ovary.

FIREWEED

Epilobium angustifolium L.

Perennial from spreading rootstocks; stem usually simple, 1 to 9 feet tall; leaves lance-shaped, entire or remotely toothed; flowers rose to purple (rarely white) borne in long terminal racemes, or some flowers in the axils of the upper leaves; fruit 2 to 3 1/2 inches long; seeds numerous, brown, each bearing a tuft of fine white hair which aid in their dispersal.

The common name "Fireweed" is derived from the fact that this species readily moves into burned areas and other disturbed sites. It is common throughout much of North America from the coast to high elevations in the mountains.

Other species of *Epilobium* may at times be considered weedy, particularly *Epilobium paniculatum* Nutt. ex Torr. & Gray (Panicle Willowweed). It is an annual species, much branched, 1 to 3 1/2 feet tall; flowers pink, the petals deeply notched, 1/8 to 1/2 inch long. This is an extremely variable species occuring in dry habitats throughout the western United States.

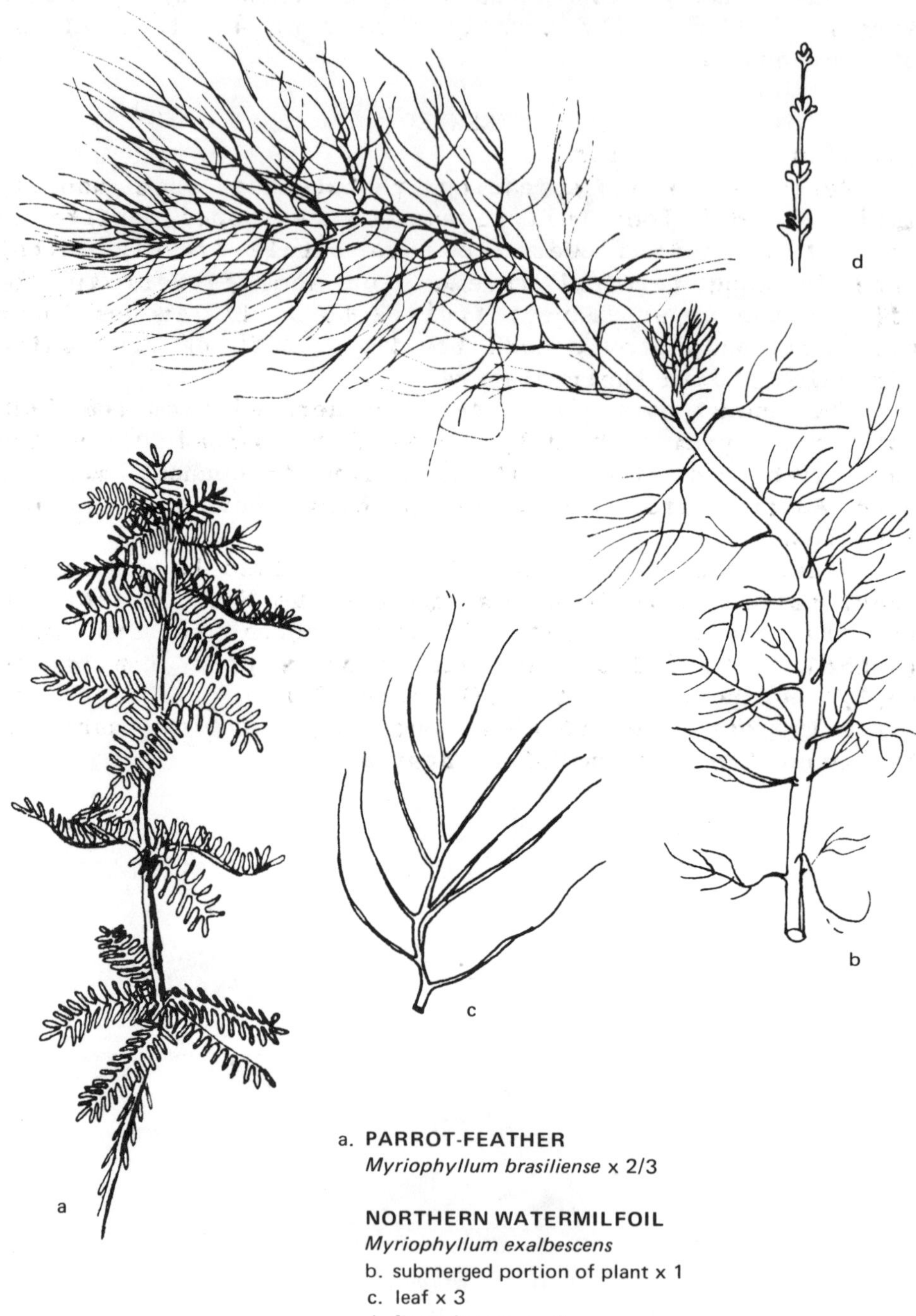

a. **PARROT-FEATHER**
Myriophyllum brasiliense x 2/3

NORTHERN WATERMILFOIL
Myriophyllum exalbescens
b. submerged portion of plant x 1
c. leaf x 3
d. flowering stem x 1

Watermilfoil Family: HALORAGACEAE

A small family of aquatic herbs with inconspicuous flowers and inferior ovaries.

PARROT-FEATHER

Myriophyllum brasiliense Camb.

Perennial; major portion of plant submerged, but with 2 to 8 inches raised above the surface of the water, at least at flowering time; stems as much as 3 feet long or more, simple or branched; leaves grayish-green, borne in whorls of 4 to 6, 3/4 to 2 inches long, divided into numerous narrow segments; flowers unisexual, borne in the axils of the upper leaves.

This South American species is a common aquarium plant and has escaped from cultivation and become sporadically established in scattered localities throughout our area. It tends to become locally abundant, covering large areas of still or slow moving water. It has become particularly troublesome along the Oregon coast in lakes, canals and irrigation ditches.

NORTHERN WATERMILFOIL

Myriophyllum exalbescens Fern.
(=*Myriophyllum spicatum* L. subsp. *exalbescens* (Fern.) Hult.)

Perennial; stems weak, often reaching 8 feet or more in length, completely submerged except at flowering time; leaves in whorls of 3 to 5, finely divided into numerous thread-like segments; flowers small, unisexual, borne in whorls in the axils of small bracts; upper bracts entire, lower bracts often toothed, length of bracts equal to or shorter than the fruit.

In ponds, lakes, irrigation ditches and streams; Alaska, Canada and the northern United States.

A similar species, *Myriophyllum verticillatum* L. (treated by some authors as *Myriophyllum spicatum* L. var. *spicatum*) is less common in our area. It has bracts which are toothed or deeply cleft and longer than the fruit.

HEDGE PARSLEY
Torilis arvensis
a. upper portion of plant x 2/3
b. root x 2/3
c. mature fruit x 4
d. basal leaf x 2/3

BUR BEAKCHERVIL
Anthriscus scandicina
e. fruits x 2/3
f. single fruit x 4
g. basal leaf x 2/3

Parsley Family: UMBELLIFERAE

To this family belong several garden vegetables and various herbs used in cooking. Examples are Celery, Parsnip, Carrot, Parsley, Anise, Caraway, and Fennel. The family also contains a number of weeds and poisonous plants.

Most members of the family have divided leaves (some exceptions) with sheathing bases, ridged stems, strong odors, and small individual flowers generally borne in umbrella-like clusters.

HEDGE PARSLEY

Torilis arvensis (Huds.) Link

Slender sparsely branching annual; stem 1 to 4 feet tall; lower leaves generally divided into 3 narrow, coarsely toothed leaflets, these and the stems covered by scattered white hairs flattened against the surface; flower small, white or pinkish, borne in several-rayed compound umbels; fruits about 1/8 inch long, short-stalked, borne in clusters, green at first, eventually purplish, densely covered by slender barbed bristles.

KNOTTED HEDGE PARSLEY

Torilis nodosa (L.) Gaertn.

Typically lower growing than the preceding (generally less than 1 1/2 feet tall); leaves more divided; fruit bristly only on the outer face, covered on the inner by low tubercles.

In comparatively recent years these two European species of Hedge Parsley have begun to invade the Far West. As yet they are generally found locally limited, but occasionally occur in abundance about farm buildings and in gardens. An oriental species, *Torilis japonica* (Houtt.) DC. (Japanese Hedge Parsley) is likewise reported, but appears to have gained no serious foothold.

BUR BEAKCHERVIL or BUR CHERVIL

Anthriscus scandicina (Weber) Mansfeld.

Annual; stem slender or sometimes stout, somewhat branching, 3/4 to 3 feet tall, plant generally minutely white-hairy when young; leaves stalked, finely divided, fern-like; flowers very small, borne in several-rayed compound umbels; fruit about 1/8 inch long, dark at maturity, beaked, covered by minute hooked bristles.

Bur Beakchervil, a native of Europe, was introduced into America as a garden herb, having long been esteemed as a salad plant. Since it seeds readily and thrives in almost any type of soil, it has naturally escaped the bounds of cultivation and is becoming established in many parts of this country. In the Northwest it is found particularly west of the Cascades, sometimes in abundance about old buildings and in farmyards. Occasionally it is reported as a garden and lawn weed.

SHEPHERD'S-NEEDLE or VENUS'-COMB

Scandix pecten-veneris L.

Slender, taprooted annual; stems 4 to 12 inches tall, usually much branched from the base; leaves finely dissected, borne on stalks nearly equalling the length of the blades, base of the stalks sheathing the stem; flowers white, borne in umbels, rays of the umbel 1 to 3, but usually paired, rays 3/8 to 1 1/2 inches long at maturity; body of the fruit 1/4 to 5/8 inch long, minutely warty, beak of the fruit slender, somewhat flattened, reaching as much as 2 3/4 inches in length at maturity.

Native of Eurasia and sparingly introduced in this country. In our area it is an occasional weed of roadsides, cultivated fields and waste places west of the Cascades, only rarely reported from east of the mountains.

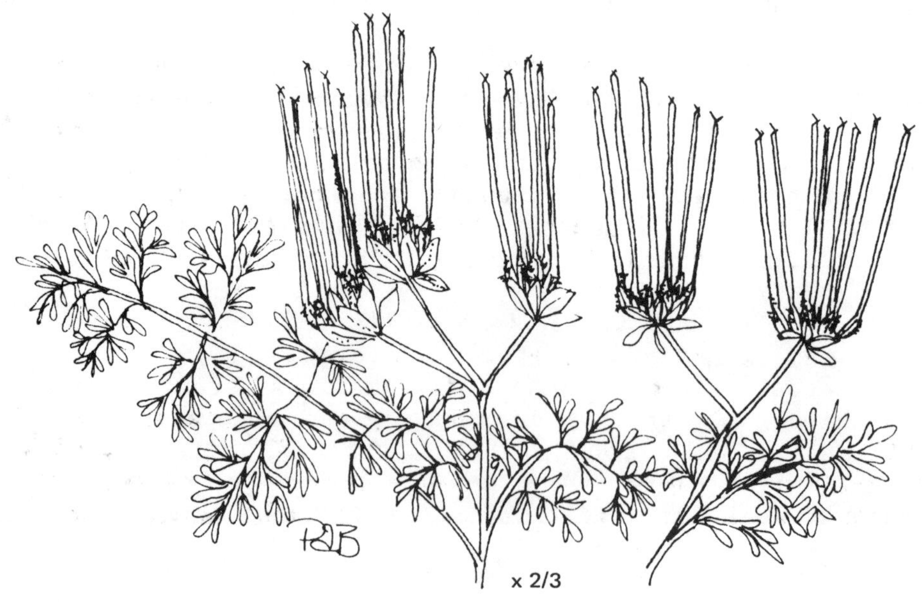

COW PARSNIP

Heracleum lanatum Michx.

Robust perennial; stems often 1 inch in diameter, 3 to 9 feet tall; leaves large, divided into 3 (rarely 5) lobed and irregularly toothed leaflets; leaflets more or less woolly, at least on the lower surface, petioles broadly sheathing; flowers in large compound umbels, white or cream-colored or tinged with pink or green; fruit winged, 1/4 to 1/2 inch long, light brown with several reddish marks running down from the apex.

This is a native species occurring in moist, shady habitats throughout much of North America. It may be troublesome along streams, fences and at the margins of pastures.

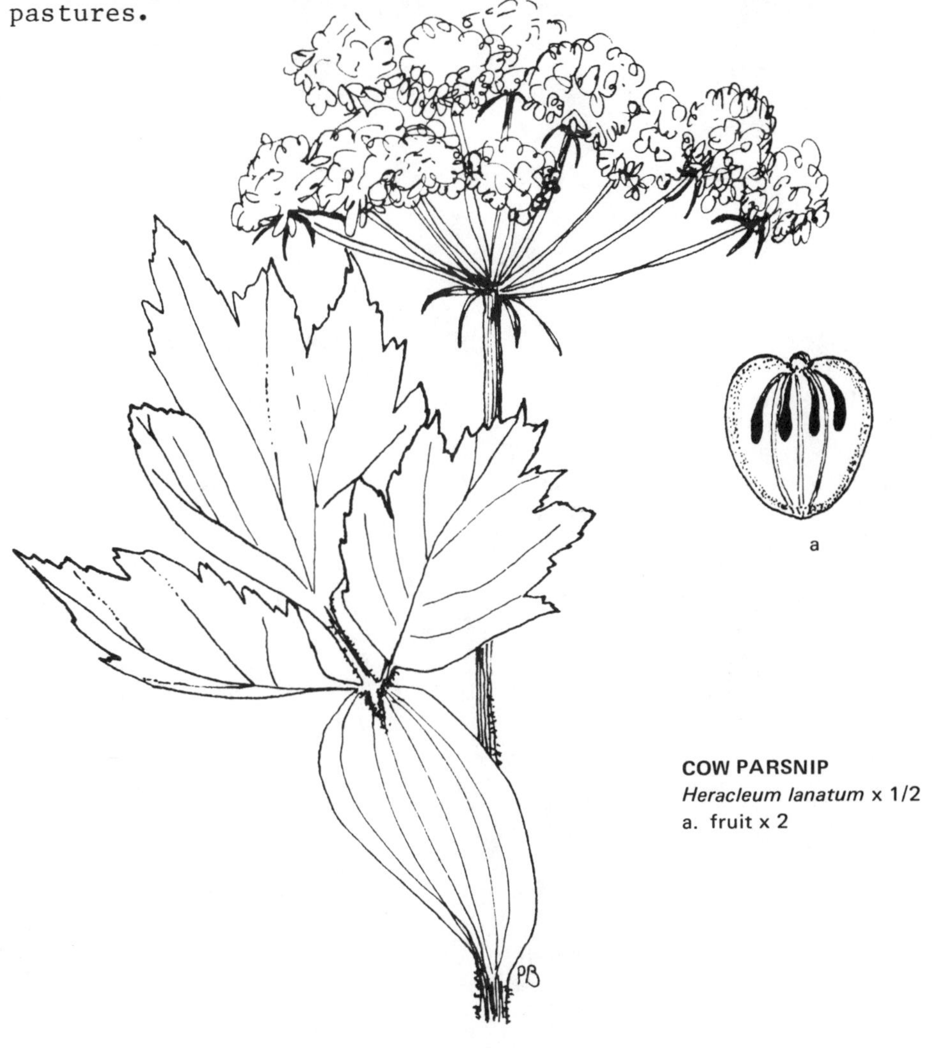

COW PARSNIP
Heracleum lanatum x 1/2
a. fruit x 2

POISON HEMLOCK
Conium maculatum x 1/2
a. fruit x 6

POISON HEMLOCK

Conium maculatum L.

Biennial herb; stem generally stout, erect, branching, hollow, ridged, usually purple-spotted especially below, 5 to 10 feet tall; leaves finely divided, fern-like; flowers white, in generally numerous compound umbels; fruit about 1/8 inch long, slightly elongated, somewhat flattened, ribbed logitudinally, the ribs knotted-wavy.

Poison Hemlock is a native of Europe where, tradition states, its poisonous properties were utilized in early days as a means of disposing of one's political or other enemies. This is said to be the plant from which the fatal dose administered to Socrates was brewed. The plant is toxic both to man and beast. That more stock-poisoning cases from this source are not reported is due in part to the fact that evidently it is not relished as food, and therefore not used when other pasturage is available. All parts of the plant are poisonous, even the underground structures. Pigs are known to have died from rooting out and eating these parts.

Strangely, in spite of its poisonous qualities, Poison Hemlock has been introduced as a garden ornamental. Its large size, dark green fern-like foliage and its abundance of lacy white umbels at flowering time, together with its ability to thrive under almost any condition of sun or shade, moisture or dryness, make it a useful background plant or filler. However, it readily escapes from cultivation and now is found in many areas as a dangerous weed of waste land, pastures, and roadsides.

Poison Hemlock is sometimes mistaken for Anise, with serious if not always fatal results. The combination of four characters which are generally observable, serves to distinguish this species from other members of the Parsley family. These are:

1. Finely divided leaves
2. A distinct mouse-like odor when the plant has been enclosed in a container for several hours or has been stored in a closed room
3. Spotted stems
4. Wavy longitudinal ribs on the fruits

For complete identification, all four characters should be used; but any one or more should arouse suspicion. The only safe rule in relation to the use of any member of this family is to know the plant thoroughly. Superficial resemblance is not sufficient ground to justify experimentation.

WESTERN WATER HEMLOCK
Cicuta douglasii
a. portion of leaf x 2/3
b. single leaflet x 1
c. longitudinal section of rootstock x 1
d. fruit x 5
e. cross-section of fruit x 5
f. portion of umbel x 2/3
g. single flower x 4

WESTERN WATER HEMLOCK

Cicuta douglasii (DC.) Coult. & Rose

Perennial; older plants with thick fleshy rootstocks, longitudinal section of the rootstock revealing cavities separated by thin plates; stem erect, more or less branching, ridged, 2 to 6 feet tall; leaves two or three times divided into narrow toothed leaflets, these 1 to 4 inches long, primary veins of the leaflets appearing to extend to the notches between the teeth; flowers small, white, borne in many-rayed compound umbels; mature fruit about 1/8 inch broad, rounded, slightly flattened, pale brown with dark vertical stripes, the two styles shorter than the fruit and turned downward.

One of our most violently poisonous plants, this species is a native of the Pacific Coast and, in Oregon, is found both east and west of the Cascades. It grows in quiet water of shallow ponds, or in swamps and marshes. In irrigated land it is common along the ditches and in areas of overflow. Often it is accessible to stock along the margins of drinking pools where the rootstocks are readily pulled out of the soggy ground.

Local names are varied. "Hemlock" is misleading in America where this name is associated with several species of forest trees. The poisonous herbs to which it is applied (both *Cicuta* and *Conium*), however, apparently have a prior right to the name since it was so used in Europe before the American trees were known.

Cicuta douglasii is likely to be confused with other members of the Parsley family, particularly with *Oenanthe sarmentosa* Presl (Water Celery) and *Sium suave* Walt. (Water Parsnip). *Cicuta* can be recognized by: (1) the chambered rootstock; (2) leaf, including the vein arrangement; and (3) fruit (see the illustration comparing leaf shapes and fruits). Only in *Cicuta* do the lateral veins of the leaflets appear to end at the notches rather than the teeth. (In reality, after branching at the notch, the veins continue along the margin of the tooth to its apex, but this point is comparatively obscure.)

Another species of *Cicuta*, commonly called Bulb-bearing Water Hemlock (*Cicuta bulbifera* L.) is only occasional in our area. It differs from Western Hemlock by the presence of bulbils in the upper leaf axils. It, also, is highly toxic.

WATER CELERY

Oenanthe sarmentosa Presl

Perennial, with thick rootstocks; plants dark green; stems spreading, branching, leafy, sometimes prostrate and rooting at the lower nodes; leaves twice divided, the leaflets broad at the base, toothed or lobed above; flowers white, in compound umbels; fruit elongated, ridged, the styles long, nearly erect.

Water Celery is a native of the Pacific Coast, and one of our most common swamp-inhabiting species, especially west of the Cascades. Since it grows under similar conditions and belongs to the same family, this plant is often confused with Water Hemlock. Its abundance about sloughs and water holes in western pastures makes it a constant source of fear to farmers. However, this plant, so far as known, is completely harmless to stock, and a few characters readily distinguish it from Water Hemlock. These identifying characters are emphasized in the illustrations and the accompanying discussions.

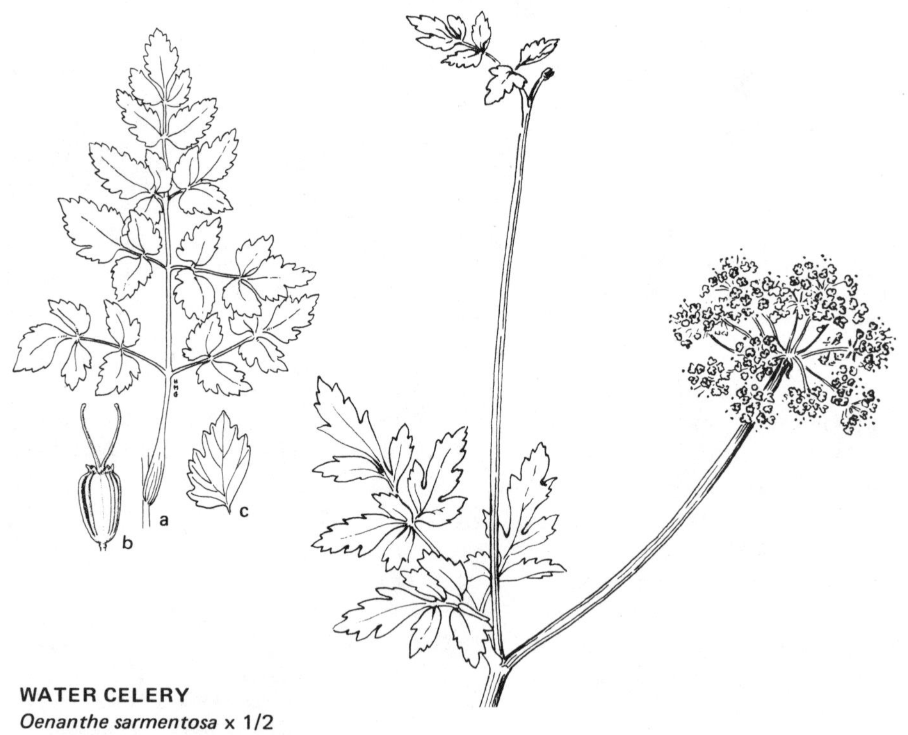

WATER CELERY
Oenanthe sarmentosa x 1/2
a. lower leaf x 1/2
b. fruit x 3
c. single leaflet x 2/3

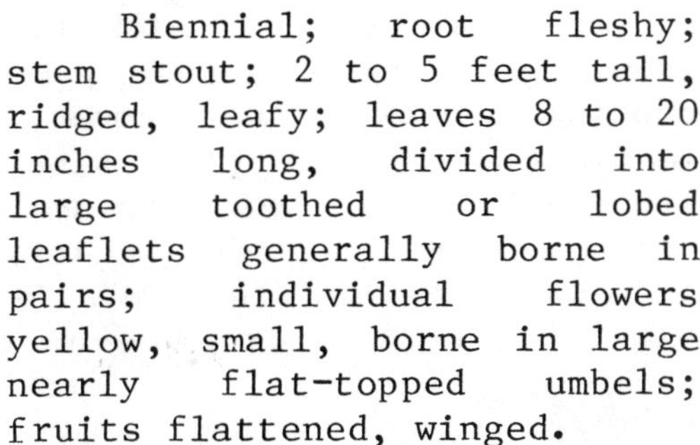

WILD PARSNIP

Pastinaca sativa L.

Biennial; root fleshy; stem stout; 2 to 5 feet tall, ridged, leafy; leaves 8 to 20 inches long, divided into large toothed or lobed leaflets generally borne in pairs; individual flowers yellow, small, borne in large nearly flat-topped umbels; fruits flattened, winged.

Pastinaca sativa is in reality the garden Parsnip, and is here included because in various regions east of the Cascades it has escaped from cultivation and "run wild," particularly in irrigated land and along the ditches.

In this process of becoming naturalized, its root has largely lost its succulence and has become somewhat woody, but the characteristic parsnip odor remains.

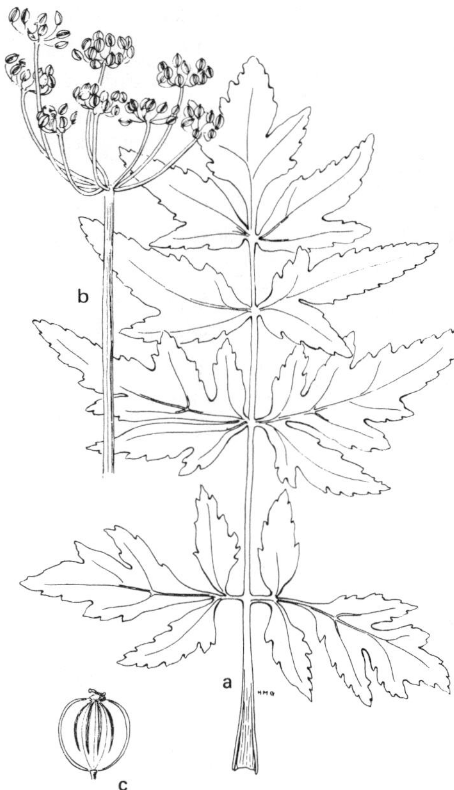

WILD PARSNIP
Pastinaca sativa
a. lower leaf x 1/3
b. umbel of fruits x 1/3
c. fruit x 2

WILD CARROT
Daucus carota x 1/2
a. fruit x 8

WILD CARROT or QUEEN ANNE'S LACE

Daucus carota L.

Biennial; tap root with distinct odor; whole plant stiff-hairy; stem branched, 1 to 3 feet tall; leaves several times divided into small toothed leaflets; flowers small, white, borne in compound flat-topped umbrella-like clusters (umbels), these umbels 2 to 4 inches in diameter, often centered by one or several minute dark purple flowers, the umbel surrounded at the base by a circle of bracts; individual fruits 1/8 to 1/6 inch long; brown at maturity, or gray after weathering, ridged, bearing rows of bristles.

Since 1900 Wild Carrot has spread from a few scattered plants to practically all roadsides, vacant lots, and many gardens west of the Cascades. In areas east of the mountians, also, it is becoming a problem. Its present abundance and prolific seeding insure a perpetual source of supply for the future, provided conditions remain unchanged.

After flowering, the "spokes" of the umbel turn somewhat inward, holding the developing fruits in a tight cluster. Gradually most of these are dislodged; but a few, tightly clutched by the inturned rays, remain throughout the winter until eventually released by weathering and general disintegration.

a. **SWAMP CANDLE**
Lysimachia terrestris x 2/3

b. **MONEYWORT**
Lysimachia nummularia x 2/3

Primrose Family: PRIMULACEAE

This family is well known from garden and greenhouse species of Primrose and Cyclamen, and from our native representatives of Shooting Star (*Dodecatheon*) and Star Flower (*Trientalis*). In the high mountains of the Pacific Northwest grow several exquisite species of native Primrose (*Primula*), while several other genera, represented in our area principally by inconspicuous species, occur. Three introduced members, one from the Eastern States and two from Europe, have become somewhat naturalized and troublesome.

SWAMP CANDLE

Lysimachia terrestris (L.) B.S.P.

Perennial; stem erect, 3/4 to 3 feet tall; leaves generally paired, narrow, not stalked but somewhat narrowed at the base, long tapering at the apex, dark green above, paler but dark-dotted below, 1 to 3 inches long; flowers borne in a long raceme, petal lobes narrow, yellow with dark red streaks.

Swamp Candle, which is a native of the Atlantic States, undoubtedly was introduced on the Pacific Coast with cranberry plants from eastern bogs. It is now found in cranberry marshes where it has become established as a pest. Its ability to creep by means of stolons renders it a constant menace and difficult to control.

MONEYWORT

Lysimachia nummularia L.

Perennial; stems 1 to 3 feet long, trailing or creeping, rooting at the nodes; leaves paired, nearly round or somewhat oblong, 1/2 to 3/4 inch long, microscopically dotted, short-stalked; flowers yellow, approximately 1 inch in diameter, borne in the leaf axils and, like the leaves, glandular-dotted.

Moneywort, whose common name is derived from the typical shape of the leaves, is a native of Europe. Its invasion of our area has occurred in comparatively recent years, but it is now abundant locally in various low areas where it has crowded out the native vegetation. In irrigated fields it could become a serious problem.

SCARLET PIMPERNEL

Anagallis arvensis L.

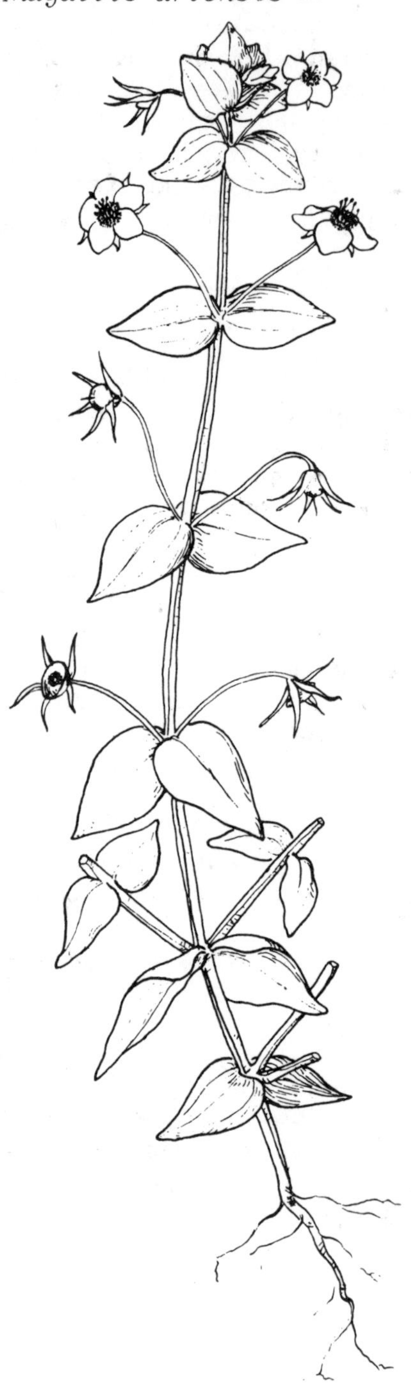

Annual; stems 4-angled, prostrate to weakly erect, up to 1 foot long; leaves up to 3/4 inch long, paired, sessile and somewhat clasping at the base, minutely glandular-dotted beneath; flowers on slender stalks in the leaf axils; petals orange or salmon-colored; fruit globose, the top separating, like a cap from the base to allow escape of the seeds.

This native of Eurasia has been introduced as a weed in temperate North America. In our area it is most common along the coast and in the interior valleys west of the Cascades, decreasing in occurrence northward.

x 1 1/4

Dogbane Family: APOCYNACEAE

Plants with milky juice. In our species, the leaves are paired, and the two fruits of each flower become long, narrow, and dry at maturity. The ornamental shrub, Oleander, belongs to this family.

BIGLEAF PERIWINKLE

Vinca major L.

Perennial; stem long, trailing, rooting at the nodes, flowering tips erect, leaves paired, broad, thick, shining, evergreen; petals twisted in the bud, spreading wheel-shaped in flower, deep blue, rounded at the apex, the flower reaching 2 inches in diameter, borne in the leaf axils; fruit 2 per flower, long, slender at maturity when present, but often not developing.

This is a well-known garden plant introduced from Europe. It makes a satisfactory ground cover where needed, but unfortunately it easily escapes bounds and may become a nuisance by encroaching on territory where it is not wanted. It is frequently found as a roadside weed or about old buildings where it has "run wild".

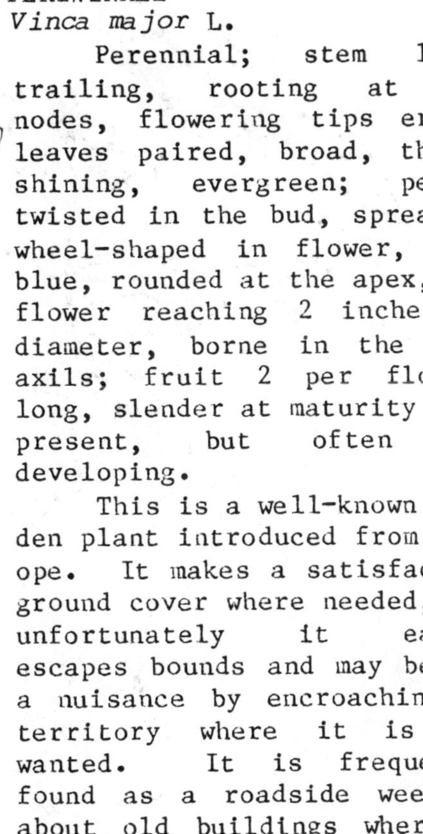

x 2/3

SPREADING DOGBANE
Apocynum androsaemifolium x 2/3
a. paired fruits from a single flower x 2/3

SPREADING DOGBANE

Apocynum androsaemifolium L.

Perennial; stem 3/4 to 2 feet tall, with spreading branches; plant generally smooth, with milky juice; leaves 1 to 2 or more inches long, dark green above, paler and hairy beneath, very short-stalked, spreading or sometimes turned downward, blades rounded at the base, narrower at the apex; flowers white or generally pale pink or pink-striped, approximately 1/4 inch long, petal tips turned backward, flowers borne in loose open clusters at ends of stems and in axils of upper leaves; fruits long (2 1/2 to 6 inches), narrow, green or dark reddish brown, a pair from each flower, each splitting longitudinally at maturity; seeds many, small, each tipped by a tuft of long hairs.

Several other species and varieties of Dogbane occur within our limits, some of which are occasionally found as weeds in gardens or orchards. Probably all are poisonous, both to man and stock, when taken in quantity, and certain species are employed medicinally.

MEXICAN WHORLED MILKWEED
Asclepias fascicularis x 1/3
a. fruits from four flowers x 1/3
b. fruit splitting open to release seeds x 1/3
c. leaf x 1/2

SHOWY MILKWEED
Asclepias speciosa x 1/3
a. fruit x 1/4
b. seed x 2/3

Milkweed Family: ASCLEPIADACEAE

This family is closely related to the Dogbane Family, and its members likewise generally contain a milky juice. The two families differ in several details, the most conspicuous being the occurrence in the Milkweed Family of stamens fused with each other and with the tip of the pistil. In the Dogbane Family the stamens are free from each other, though sometimes lightly attached to the tip of the pistil.

MEXICAN WHORLED MILKWEED or NARROW-LEAVED MILKWEED

Asclepias fascicularis Dcne.

Perennial; creeping underground and sending up new plants at intervals; stems 2 to 4 feet tall, smooth, dull green, or often purplish; leaves narrow, 1 to 4 inches long, sometimes folded along the midrib, generally 3 or more leaves at each joint, erect or spreading or often curved downward; flowers pale purplish or greenish, borne in small long-stalked umbrella-like clusters arising at the leaf axils; fruit slender, brown; seeds many in each fruit, each seed with a tuft of long hairs.

A native of western America, perhaps first known in Mexico, this species is now found in most of the Western States. In Oregon it occurs on both sides of the Cascades in dry or moist areas, pastures, fields and along irrigation ditches. It is poisonous to stock, though apparently in the green state it is little relished and eaten only for want of better pasture. When dry, as in hay, it loses its disagreeable odor, thus escaping detection although its poisonous qualities remain, and animals then eat it readily.

SHOWY MILKWEED

Asclepias speciosa Torr.

Perennial from spreading rootstocks; stems 1 to 3 feet tall, gray-woolly at least below; leaves opposite, 3 to 8 inches long and up to 4 inches wide; flowers in umbrella-like clusters, pink to reddish-purple; fruits woolly and with soft prickles.

Usually in sandy or gravelly, often moist, soil. This species occurs on both sides of the Cascades, but is more common eastward.

Showy Milkweed is reported to be toxic but less dangerous than Mexican Whorled Milkweed.

FIELD DODDER

Cuscuta pentagona var. *calycina*

a. Field Dodder parasitic on clover x 3/4

b. flower x 4

c. seeds x 6

Morningglory Family: CONVOLVULACEAE

Most members of this family are trailers or climbers, the larger forms generally bearing flaring trumpet-shaped flowers. Here belong Sweet Potato and the so-called Wooden Rose commonly grown in Hawaii. In this family are found, also, certain inconspicuous species bearing minute flowers.

Generally considered as a branch of the family is the group of parasites known as Dodder, or these may be treated as belonging to a segregate family: CUSCUTACEAE. Some of these are native, some introduced, but all are harmful to whatever plants they attack as "hosts."

FIELD DODDER

Cuscuta pentagona Engelm. var. *calycina* Engelm.
(=*Cuscuta campestris* Yunk.)

Annual; parasitic, yellow or reddish, twining about other plants and obtaining nourishment from them by means of "suckers;" flowers in small dense clusters; seeds about 1/16 inch long.

Ten or more species of dodder are known to occur in our area, all of them parasitic, all with slender weak, golden-yellow or reddish stems often borne in such dense masses that at maturity their presence in a field is easily observable from a distance. All our species derive their nourishment from various host plants such as clover, alfalfa, grasses, goldenrod, and others equally varied.

Dodder seeds are small and in certain species not easily distinguished from those of alfalfa, though the tendency of some dodder seeds to cling together in small clusters is a slightly helpful recognition point. The seeds germinate on the ground, sending a short root downward into the soil, and a thread-like stem upward, the latter to swing about until it comes in contact with an object about which it can twine. If this object is a green plant, the dodder soon develops "suckers" which penetrate the host and absorb food. The ground attachment disappears, and the parasite now lives wholly upon its host. Branching and rebranching, the slender pliable stems reaching out from plant to plant, continually forming new "suckers," dodder may spread rapidly across a field, often exhausting the host plants as it goes. The seeds are comparatively long-lived, and delayed germination may account for reappearance of the pest in a given field after several years.

FIELD BINDWEED OR FIELD MORNINGGLORY

Convolvulus arvensis L.

Perennial by rootstocks and an extensive root system; stems prostrate or twining; leaves more or less arrow-shaped, slender-stalked; flowers stalked, borne singly or in pairs in the leaf axils; petals fused into a funnel-shaped tube, pink or white, approximately 1 inch in diameter, the flower stalk with 2 minute bracts borne considerably below the flower; fruit broad, abruptly pointed, typically containing 4 seeds; seed about 1/8 inch long, dark brown, roughened.

One of the most troublesome of introduced Old World weeds, particularly in orchards, gardens, and grain and mint fields. When growing with other plants it is a twining vine, but if no support is convenient, it lies prostrate on the ground. The stems are slender and rarely straight since they twine about everything in reach, coiling about each other if no other support is present.

The characteristic which makes of this attractive harmless-appearing vine a noxious weed is its underground system which is perennial, creeping, and able to send up new plants from any node. These rootstocks and roots, under typical conditions, are filled with starch which has been manufactured by the leaves and stored below ground for future use. It is these reserves of food, found in enormous quantities in mature plants and constantly replenished as the store is drawn upon by growing points, which make the weed difficult to combat. Ordinary cultivation improves the conditions under which it grows, and stimulates it to further growth. Every smallest portion of green leaf or stem allowed to appear above ground adds to the starch supply below, the roots grow deeper, the rootstocks spead more broadly, and a whole field may become infested from a single plant, both vegetatively and by means of abundant seed production.

This Morningglory will grow in practically any soil, but the deeper and richer the soil, the ranker the growth. It was estimated by careful weighings conducted some years ago on the University of California farm at Davis, that ordinary root growth may reach from 2 1/2 to 5 tons per acre on heavily infested land. The quantities of food stored within these tons of root and rhizome may enable the plants to live for many months with no shoot appearance above ground. The enormous root system also robs the land of moisture and food substance during the growing season, and thus sometimes destroys even young orchard trees or makes old orchards valueless.

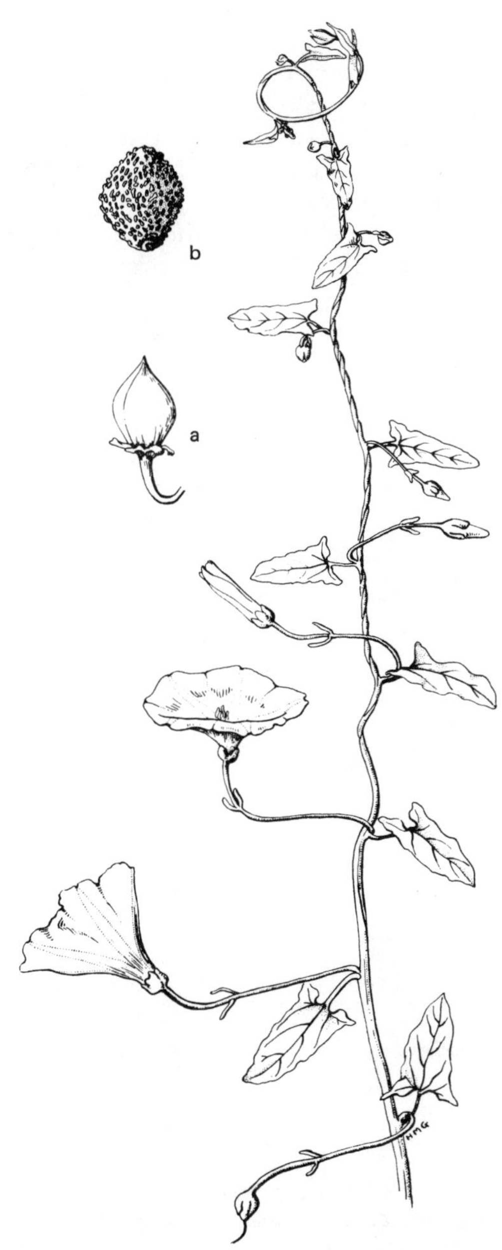

FIELD BINDWEED
Convolvulus arvensis x 2/3
a. fruit x 2
b. seed x 4

HEDGE BINDWEED or LARGE WILD MORNINGGLORY

Convolvulus sepium L.

Perennial from elongated rootstocks; stems reaching 9 feet long, climbing over other vegetation; leaf blades 1 1/2 to 5 inches long, generally broadly arrow-shaped, long stalked; flowers trumpet-shaped, white to deep pink, 1 1/2 to 3 inches long; fruit nearly globose, about 3/8 inch long, splitting open at maturity to release the 2 to 4 seeds; seeds gray to black, minutely roughened.

This species is a native of the eastern United States. It is only sparingly established along the West Coast where it becomes a serious pest, often densely covering banks and roadside shrubbery; when it gets into ornamental plantings, it is difficult to control.

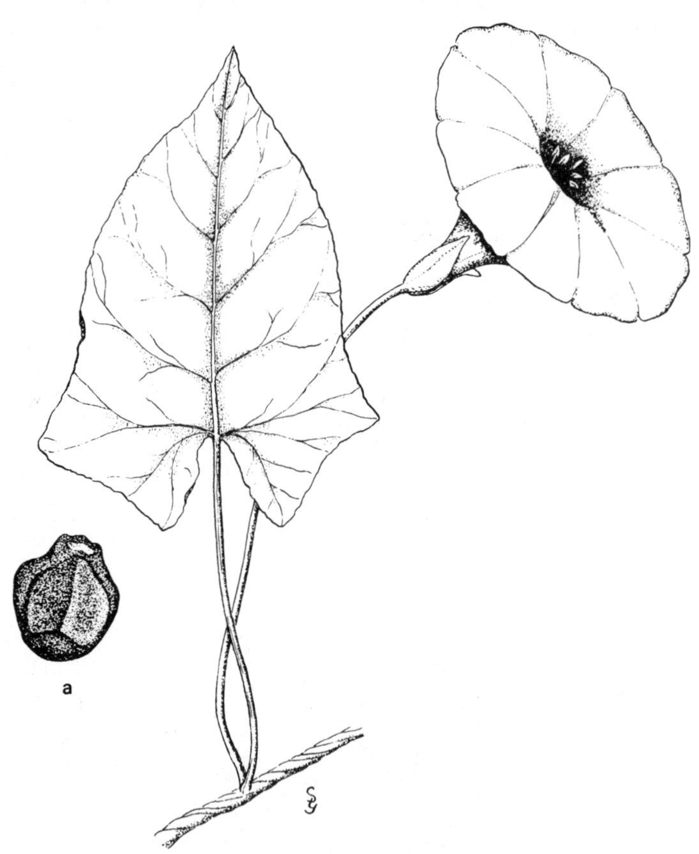

HEDGE BINDWEED
Convolvulus sepium x 4/5
a. seed x 3

Waterleaf Family: HYDROPHYLLACEAE

This family, whose members resemble in several respects those of the Borage Family, is distinguished from it by the fruits which split at maturity to release the seeds, rather than separating into 4 nutlets.

Several ornamentals and a few weedy species belong here; but only the following is generally sufficiently abundant in a field to be significant.

WHITE BABY-BLUE-EYES

Nemophila menziesii H. & A. var. *atomaria* (F. & M.) Chand.

Annual; stems weakly erect or spreading, 2 to 8 inches high, branching widely; leaves 1/2 to 1 inch long, lobed, hairy; flowers stalked, 1 inch or less in diameter, white (rarely blue or tinged with blue), lined and dotted with black or dark blue; fruit nearly globose; seed 1/12 inch long, dark, warty.

The only justification for inclusion of this species in a weed manual is its abundance, at times, in early spring when it spectacularly forms white drifts of blossom over the previous year's cultivated fields. Its season is so short and the entire plant disappears so early that little competition occurs between it and the normal crop. In the southern part of our range the variety gives way to typical *Nemophila menziesii* with showy blue flowers.

x 2/3

FIELD GILIA
Gilia capitata x 2/3
a. seeds x 6
b. lower leaf x 2/3

Phlox Family: POLEMONIACEAE

Better known by its garden ornamentals than its weeds, this family nevertheless contains a few species which merit attention as potential weeds. Members of the family in general have trumpet-shaped or bell-shaped flowers, often brightly colored and, in many species, clustered at the ends of branches or borne in the leaf axils.

FIELD GILIA OR GILLYFLOWER

Gilia capitata Sims.

Annual; generally erect, nearly simple or sometimes widely branched from the base upward, 1/6 to 3 or more feet tall; first leaves crowded at the base of the stem, later leaves borne alternately along the stem, leaves divided into narrow segments; flowers pale to dark blue or white, borne in dense head-like clusters.

It is a common plant of the Pacific slopes, and found also in the Cascades and eastward. It is a familiar sight in grain fields of the western valleys, but is likewise at home in mossy crevices of wet cliffs or on dry rocky hillsides.

ANNUAL POLEMONIUM

Polemonium micranthum Benth.
(= *Polemoniella micrantha* (Benth.) Heller)

Weak-stemmed annual; leaves 3/4 to 1 1/2 inches long, composed of 7 to 15 leaflets; flowers appearing opposite the leaves; petals white, united at base into a yellowish tube, more or less concealed by the sepals; fruit splitting at maturity to release the seeds.

This native species is common in moist ground throughout much of our area east of the Cascades. It is occasional as a weed in cultivated areas.

x 1/2

SKUNKWEED GILIA

Navarretia squarrosa (Esch.) H. & A.

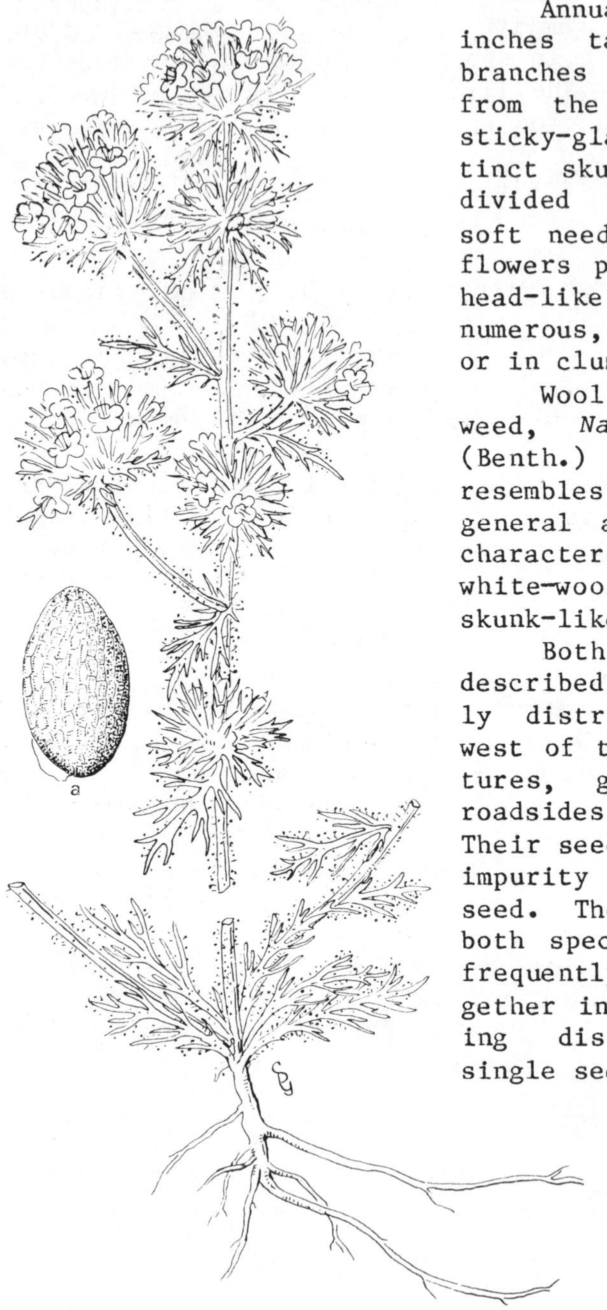

Annual; 3 to rarely 18 inches tall, erect, or the branches sometimes spreading from the base; entire plant sticky-glandular, with a distinct skunk-like odor; leaves divided into narrow rather soft needle-pointed segments; flowers pale blue, in compact head-like clusters; seeds numerous, falling separately or in clumps.

Woolly Gilia or Needleweed, *Navarretia intertexta* (Benth.) Hook., considerably resembles Skunkweed in size, general appearance, and seed characters, but its heads are white-woolly and it lacks the skunk-like odor.

Both species here described are native and widely distributed, particularly west of the Cascades, in pastures, grain fields, along roadsides, and in gardens. Their seeds may be found as an impurity in grain and grass seed. The individual seeds of both species are minute, but frequently they are glued together in clusters, these being distributed with the single seeds.

SKUNKWEED GILIA
Navarretia squarrosa x 1
a. seed x 24

Borage Family: BORAGINACEAE

This family is best known by Forget-me-not, Heliotrope, and other species of favorite garden plants. Many members of the family have flowers suggesting those of Forget-me-not, both in shape and in color succession as the flower matures. The fruit is 4-lobed, separating at maturity into 4 one-seeded nutlets. In some species these nutlets are smooth, but in others they are ornamented by spines or hooks or prickles which disagreeably adhere to clothing, or the wool or hair of animals, and which may cause injury to the lips and nostrils of grazing stock. Some of the species are poisonous.

COAST FIDDLENECK

Amsinckia intermedia Fisch. & Mey.

Annual; stem erect, clothed with stiff bristles and sometimes also with fine white hairs; leaves elongated, those at the base long-stalked, the upper leaves stalkless, broadest near the base, covered by stiff bristles; flowers yellow or orange, borne on at first incurved but gradually straightening and lengthening axes; fruit divided into 4 nutlets, each nutlet bearing an enclosed seed; nutlets slightly curved, keeled on the back, warty.

A common plant on both sides of the Cascades, and sometimes an annoying weed in cultivated fields.

Other species of *Amsinckia* occurring within our area include: *Amsinckia lycopsoides* Lehm. ex Fisch. & Mey. (TARWEED FIDDLENECK) which is generally more sprawling in habit than the preceding species; it has broader leaves and shorter flower clusters; it occurs mainly, but not entirely east of the Cascades; *Amsinckia menziesii* (Lehm.) Nels. & Macbr. (SMALL-FLOWERED FIDDLENECK) has pale yellow flowers 1/8 to 1/4 inch long and has thin, broad leaves; it occurs on both sides of the Cascades, often in moister sites than the other species; *Amsinckia retrorsa* Suksd. (RIGID FIDDLENECK) is similar to *Amsinckia menziesii* but with both spreading stiff hairs and fine appressed hairs on the stem; *Amsinckia tessellata* Gray (WESTERN FIDDLENECK) is distinguished from other species by having a reduced number of sepals (2, 3, or 4), these being very unequal in width; it occurs chiefly east of the Cascades.

COAST FIDDLENECK
Amsinckia intermedia x 4/5
a. nutlet x 10

POPCORN FLOWER or SCORPION GRASS

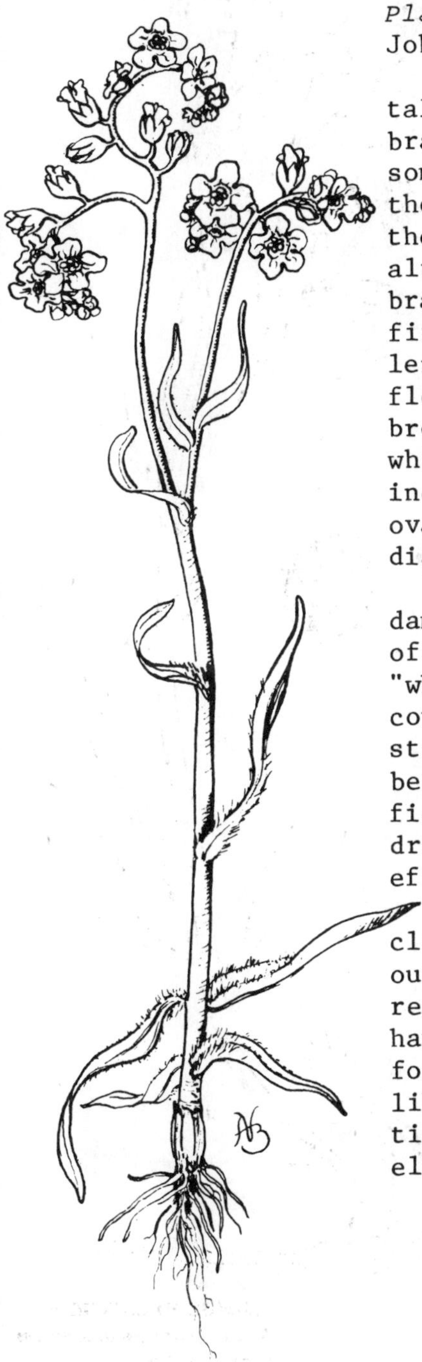

Plagiobothrys figuratus (Piper) Johnst.

Annual; plant 1/3 to 1 1/2 feet tall, slender or stout, simple or branched, entire plant generally somewhat stiff-hairy; leaves narrow, those near the base borne in pairs, the upper leaves principally alternate; flowers borne in branching clusters, the branches at first tightly inrolled, but lengthening and straightening as the flowers mature; hairs of flower-bud brown; flower forget-me-not shaped, white with yellow center, about 1/4 inch in diameter; nutlets broadly ovate, approximately 1/16 inch in diameter, with wrinkled surface.

This native species grows on damp ground west of the Cascades, often occurring so abundantly on low "white" soil that these areas appear covered by drifted snow and contrast strikingly with higher ground or better-drained soil within the same field. It is more noted as a soil-drainage indicator than for harmful effects on crops.

Many species of this and closely related forms occur within our boundaries, some inhabiting arid regions east of the Cascades. All have flowers suggesting white forget-me-not and borne in raceme-like arrangements which at first are tightly curved but which gradually elongate.

x 1

a-b. **HOUNDSTONGUE**
Cynoglossum officinale
a. portion of inflorescence x 1/2
b. developing nutlets x 3

c. **CORN GROMWELL**
Lithospermum arvense x 1/2

d-e. **SEASIDE HELIOTROPE**
Heliotropium curvassavicum
d. plant x 4/5
e. nutlets x 10

HOUNDSTONGUE

Cynoglossum offincinale L.

Biennial; stems 1 to 4 feet tall; leaves entire, lanceolate, the lower long-stalked, the upper sessile, rough hairy; flowers reddish-purple; fruit divided into 1 to 4 nutlets; nutlets prickly.

Native of Europe, well established in disturbed habitats, especially along roadsides, throughout our area.

CORN GROMWELL

Lithospermum arvense L.

Annual; stems simple or few branched, 1/4 to 2 1/2 feet tall; lower leaves early withering, the others lanceolate and sessile or nearly so; flowers white to bluish-white, borne in the axils of the reduced upper leaves; fruit divided into 4 nutlets, these gray-brown, ridged on one side, pitted and sometimes warty.

This European species is widely distributed throughout the United States; in our area it is found mainly east of the Cascades in fields and along roadsides.

SEASIDE HELIOTROPE

Heliotropium curassavicum L. var. *obovatum* DC.

Perennial; plant smooth, bluish-green; stems 1/4 to 3 feet long, branching, prostrate or weakly ascending, leafy; leaves 1 to 2 inches long, narrowed at the base, broadest near the apex; flowers white or bluish with yellow center, borne in narrow, paired, one-sided clusters which are first curved inward, later elongating and straightening; nutlets very small, light brown and roughened.

A common weed in saline places, but also spreading to irrigated land. In our area this species occurs only east of the Cascades. According to Weeds of California (Robbins, Bellue, and Ball), the nutlets are often common in alfalfa seed and may be confused with the seeds of Dodder. The significance of this is apparent and may account for the name Devilweed sometimes applied to the plant.

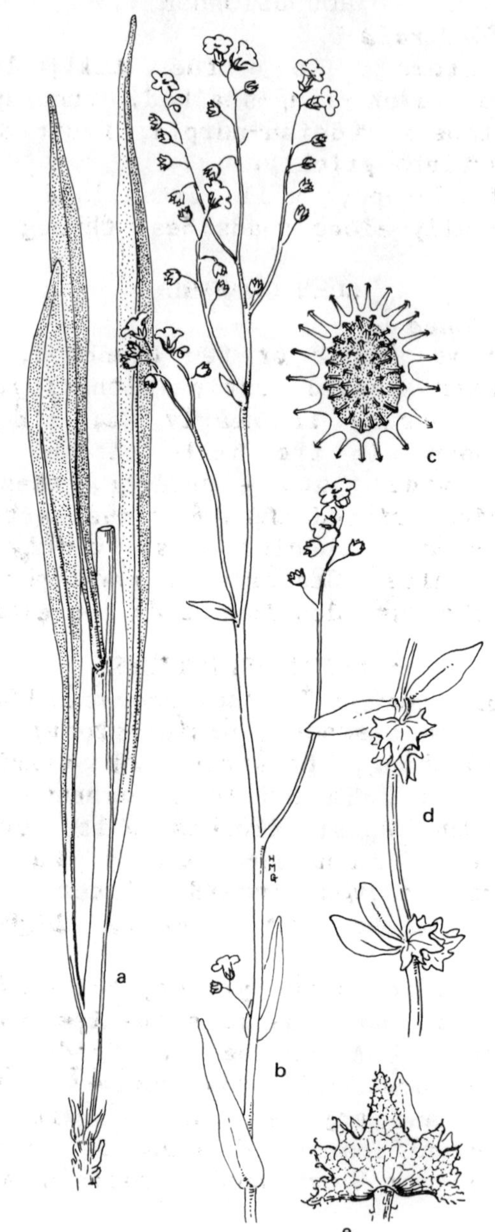

STICKWEED
Hackelia diffusa
a. lower portion of plant x 2/3
b. inflorescence x 2/3
c. nutlet x 4

MADWORT
Asperugo procumbens
d. portion of stem x 2/3
e. enlarged calyx x 1 1/2

STICKWEED

Hackelia diffusa (Lehm.) Johnst.

Biennial or perennial; stems numerous, weak, leafy; leaves narrow, lower leaves long-stalked, upper leaves stalkless, broadest at the base; flowers white or blue, forget-me-not shaped, borne in a loose branching arrangement; nutlets bristly.

A rather common species in dry areas east of the Cascades.

MADWORT or CATCHWEED

Asperugo procumbens L.

Biennial or perennial; branches low-spreading or weakly ascending, 1 to 2 feet long; leaves alternate on the stem, narrow, 1 to 4 inches long, harsh-bristly; flowers blue, borne in the upper leaf axils, short-stalked, the stalks eventually turning downward; calyx enlarging after flowering, forming a conspicuous collar about the nutlets.

An introduction from Europe, Madwort has become quite widely distributed east of the Cascades. It is a rough, disagreeable weed of cultivated fields and of roadsides.

a. **PEPPERMINT**
Mentha piperita x 3/4

b. **FIELD MINT**
Mentha arvensis x 3/4

c. **SPEARMINT**
Mentha spicata x 3/4

Mint Family: LABIATAE

Members of this well-known family are generally recognizable by 4-angled stems, paired leaves, pungent odor (a few exceptions), more or less irregular flowers, and the fruit separating into 4 nutlets at maturity.

The true mints (*Mentha*) are represented in the Pacific Northwest by one native species (*Mentha arvensis* L.) and several which have been introduced from Europe. Since most of these mints grow in wet places, they are of no consequence as weeds except in low pastures or fields. Where locally abundant, however, Peppermint, Spearmint, and Field Mint are so conspicuous and so frequently reported as to merit inclusion here. Other species occasionally occurring as escapes from cultivation are Bergamot Mint (*M. citrata* Ehrh.), Pennyroyal (*M. pulegium* L.), and several others more rarely encountered.

PEPPERMINT

Mentha piperita L.

Perennial; creeping by rootstocks and stolons; stem erect, 1 1/2 to 3 feet tall, 4-angled, often purplish; leaves widely spreading outward or downward, short-stalked, 1 to 2 1/2 inches long, blade generally broadest near the base or center, pointed or rounded at the apex, coarsely toothed, sometimes folded along the midrib; flowers lavender, borne in thick spike-like clusters at the apex of the stem and in the upper leaf axils.

A native of Europe, Peppermint long ago was introduced into our area but soon escaped cultivation and became established in marshes and lowlands west of the Cascades.

FIELD MINT

Mentha arvensis L.

Perennial; spreading by slender rootstocks; stem 1 to 2 1/2 feet tall; leaves short-stalked, blades broadest at the base, toothed; flowers pinkish to bluish, borne in dense clusters in the leaf axils.

This species, which is a native, is represented by various varieties or subspecies. It occurs in wet ground and is known practically throughout our area.

SPEARMINT

Mentha spicata L.

Perennial; creeping by stolons and/or rootstocks; distinguished from Peppermint by generally more slender, longer-pointed and more conspicuously veiny leaves, these inclined to be less widely spreading and not at all or scarcely stalked; flower spikes typically longer and more slender than those of Peppermint.

This species, likewise, is of European origin, and probably was introduced in the same manner. It also grows in damp ground and may be found in low pastures.

HENBIT OR DEADNETTLE

Lamium amplexicaule L.

Annual; stem much branched at the base, spreading or weakly erect, 4-angled, bearing few pairs of leaves; leaf blades roundish, coarsely toothed or lobed, the lower leaf blades stalked, more or less heart-shaped at the base, the upper leaves without stalks, closely clasping the stem, the pairs of upper leaves sometimes crowded above; flowers small, pink to purple and white, borne in dense clusters in the upper leaf axils, in bud the upper petal lobe drawn like a furry cap over the other flower parts; nutlets elongated, narrowed at the base, somewhat mottled with white.

This species, which is a common weed of gardens and cultivated fields west of the Cascades, has come to us from Europe. It generally appears with the first warm days of spring, or it frequently is a winter annual. Its seeds germinate in late summer, the new plants soon reach vegetative maturity and are ready to burst into flower with the earliest favorable conditions.

HENBIT
Lamium amplexicaule x 1

a. flower x 4
b. nutlet x 14

RED DEADNETTLE

Lamium purpureum L.

Annual; stem usually branched from the base, with few pairs of leaves below; leaves ovate, toothed, broadly or narrowly pointed at the apex, more or less heart-shaped at the base, all leaves stalked, the upper pairs crowded near the ends of the stems, purplish, generally pointing downward; the small purple to pink flowers borne in their axils are less conspicuous than the colored leaves.

This weed, which is a native of Europe, often forms a showy ground cover in early spring over untilled gardens, orchards, and fields. Its luxuriant tangled growth in low gardens renders cultivation difficult.

x 2/3

GROUND IVY or CREEPING CHARLIE

Glechoma hederacea L.

Perennial; stems weak, often creeping and rooting at the nodes; leaves aromatic when crushed, paired, rounded, heart-shaped, the margins bluntly toothed, at least the lower leaves long-stalked; flowers 2-lipped, the upper lip 2-toothed, the lower lip with a wide central lobe and 2 lateral lobes, bluish-purple and spotted within; fruit of four nutlets.

Native of Eurasia, introduced into this county for use as a ground cover. In some areas it proves too aggressive, escapes from cultivation and becomes established in flower borders and lawns where its control is difficult.

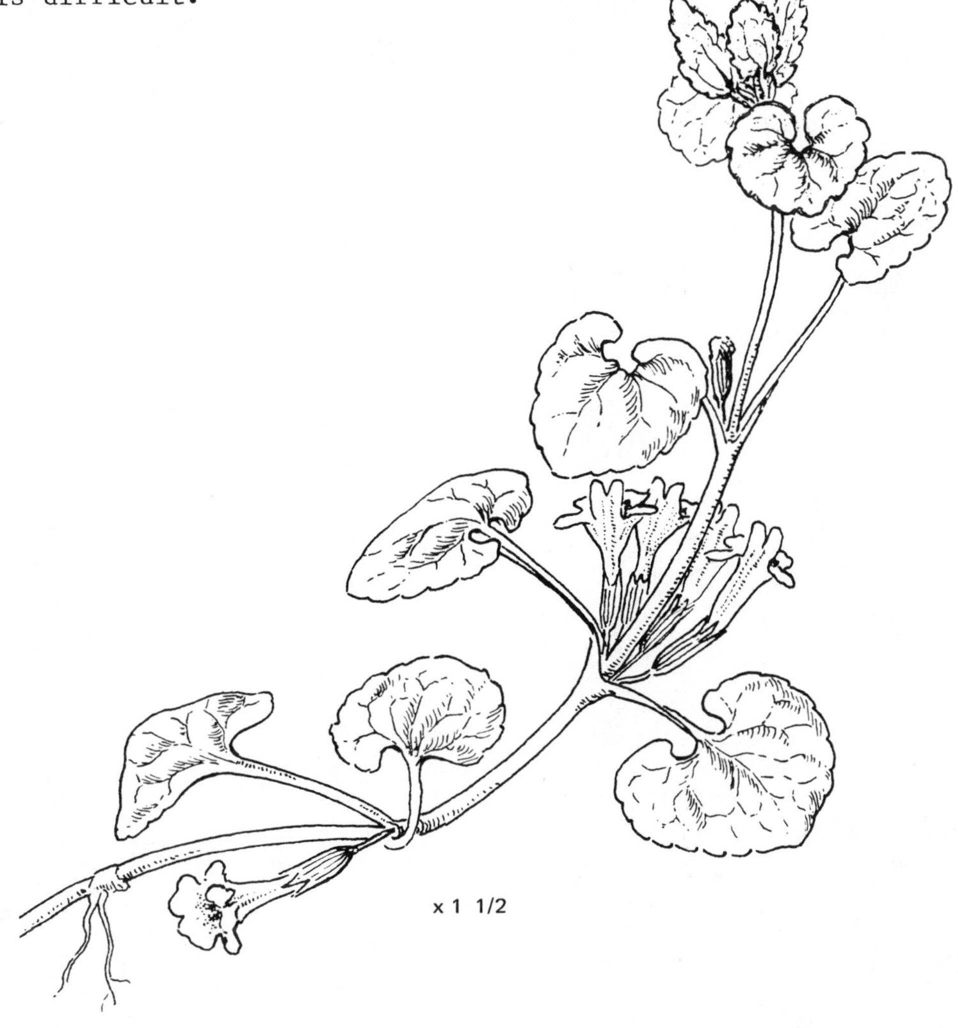

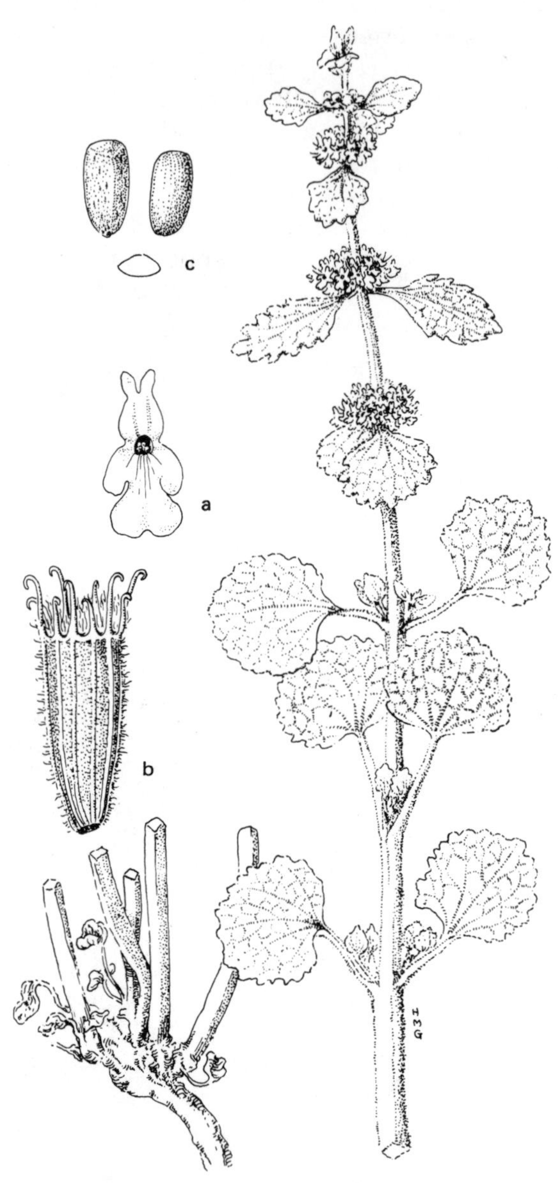

WHITE HOREHOUND
Marrubium vulgare x 2/3
a. flower x 6
b. mature calyx x 6
c. dorsal, ventral and cross-sectional views of nutlets x 6

WHITE HOREHOUND

Marrubium vulgare L.

Perennial; stem 1 to 2 1/2 feet tall, white-woolly, single or clustered from a somewhat woody base, erect; leaves paired at each joint, thickish, generally white-woolly at least beneath, stalked, the blades roundish, coarsely toothed, wrinkled-veiny; flowers small, white, borne in dense clusters in the leaf axils; fruit 4-lobed, 4-seeded, but remaining enclosed in the 10-toothed calyx, each tooth of the latter ending in a hooked spine.

This native of Europe was probably originally introduced into the United States as a garden herb, but has escaped cultivation and become widely distributed. It is particularly abundant east of the Cascades, occurring as a common inhabitant of dry areas but becoming adapted also to garden conditions. The hooked spines which enclose the fruit cling to the wool or hair of animals or to clothing, and thus are carried to new areas. Later the 4 lobes of the fruit separate, and though each nutlet produces only a single seed, the enormous number of flowers borne by each plant insures perpetuation and increase.

MEDITERRANEAN SAGE

Salvia aethiopis L.

Biennial; stem stout, 2 or 3 or more feet tall, very white-woolly; leaves at first felt-like, lower leaves stalked, blades 1/3 to 1 foot long, broad, coarsely and sharply toothed and lobed; upper leaves smaller, becoming stalkless above and clasping the stem, the upper face of all leaves sometimes losing the felt-like covering and revealing a greenish wrinkled surface; flowers small, yellowish white, borne in woolly clusters on a profusely branched arrangement at the ends of stems; nutlets smooth with dark veining.

As indicated by its common and scientific names, this plant is a native of Africa whence by some means, probably in contaminated imported alfalfa seed, it has reached western America. In Oregon it occurs east of the Cascades. Its enormous seed crop and its ability to cope with seemingly any environmental condition render its presence formidable, and in recent years it has been spreading rapidly.

The first season of its growth it produces a rosette of large grayish-felty leaves and stores food for its second season, at which time it sends up an intricately branched stem bearing numerous flowers. Each flower typically produces four nutlets, each enclosing a seed, and thousands of these may be borne by a single plant in one season. The germination rate is high, and young plants thickly cover ground which has been exposed to seeding.

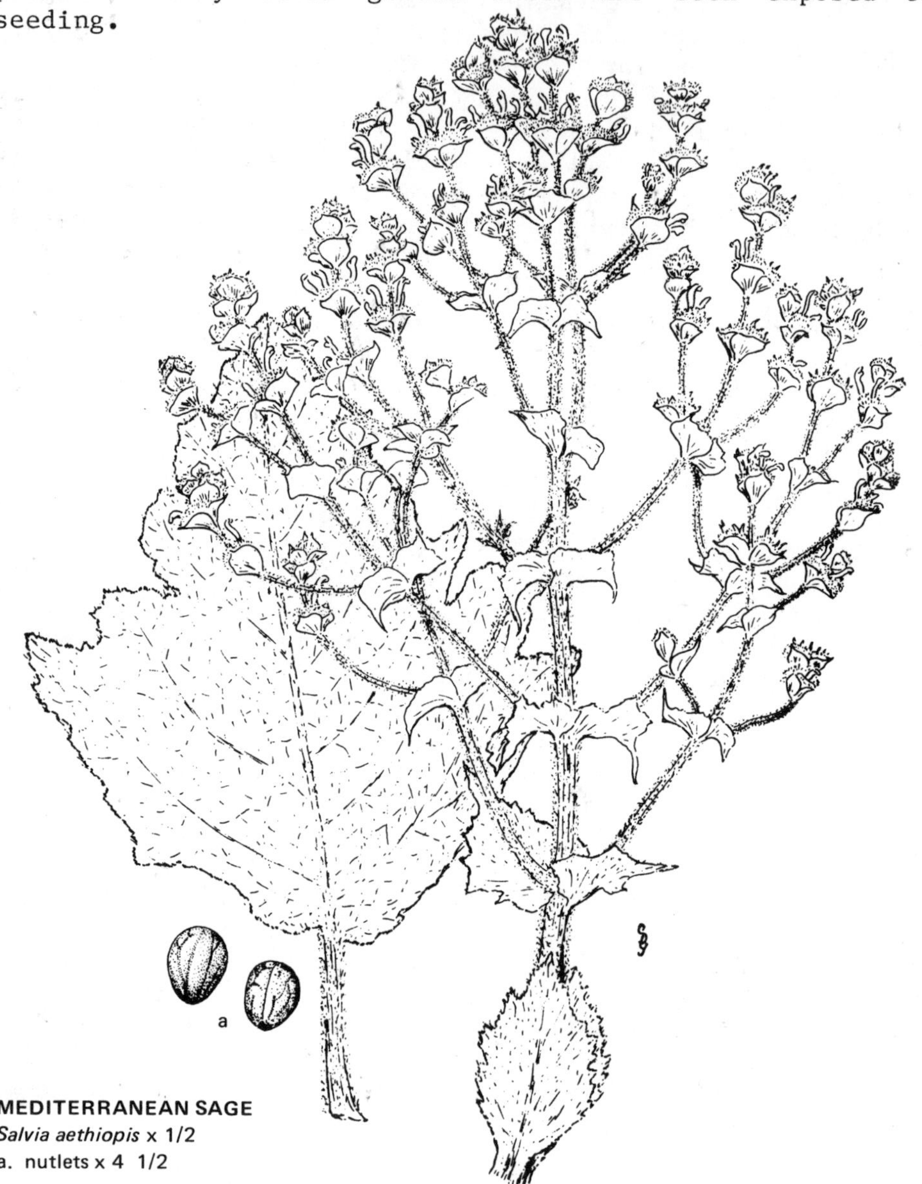

MEDITERRANEAN SAGE
Salvia aethiopis x 1/2
a. nutlets x 4 1/2

x 1

HEAL-ALL

Prunella vulgaris L.

Perennial; stem solitary or sometimes several from the base, 1/4 to 2 feet long, often horizontal below, erect above; leaves stalked, blades rounded or squared at the base, margins slightly toothed; flowers red-purple or paler, borne in a thick dense spike at apex of the stem; nutlets smooth.

Heal-all is a European plant which has been known in America probably as long as the white man. It is common in waste land, pastures, meadows, and fields. As a pest, it is best known in lawns where it spreads low on the surface, forming conspicuous patches which at times take on a purplish hue. In the open, where conditions are favorable, the flowering spikes are large, showy, and borne on long erect stalks, but in lawns, subjected to constant mowing, its growth is compact and close to the ground, even its spikes being reduced in size and more or less hidden.

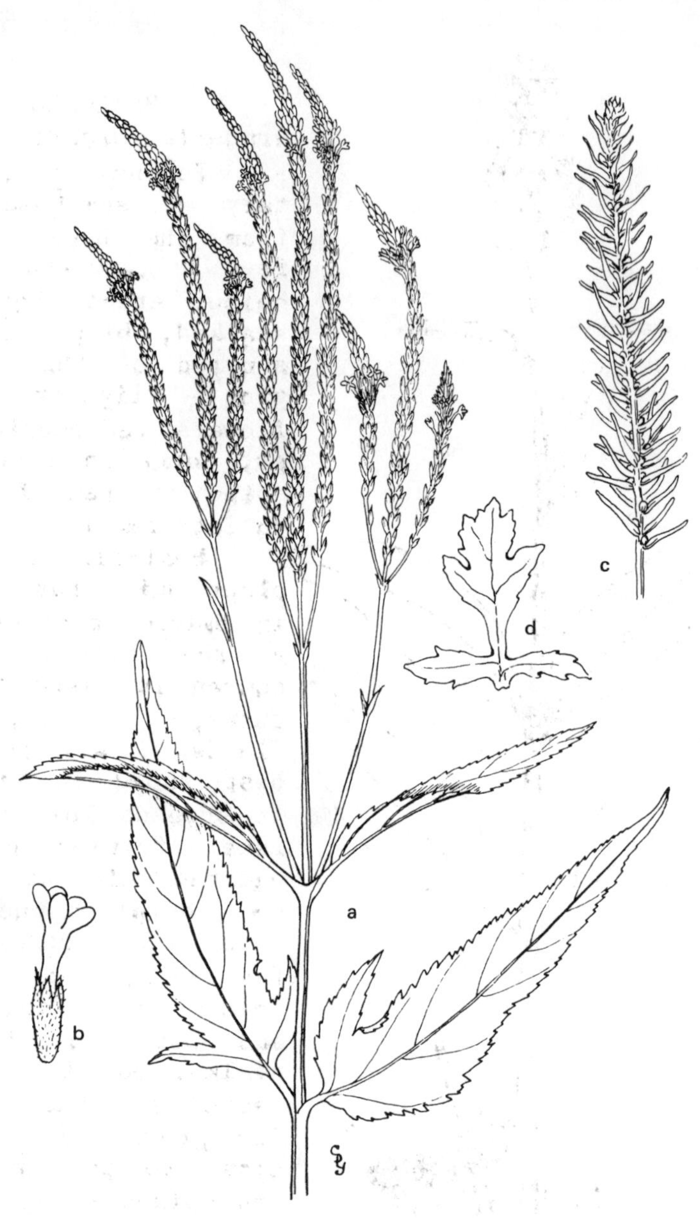

BLUE VERBENA
Verbena hastata
a. upper portion of plant x 2/3
b. flower x 4

PROSTRATE VERVAIN
Verbena bracteata
c. portion of inflorescence x 2/3
d. leaf x 2/3

Verbena Family: VERBENACEAE

Two species of *Verbena*, closely related to garden forms but less showy, are frequently reported as weeds. Both have paired leaves, flowers borne in long spikes, and fruit separating into 4 nutlets at maturity.

BLUE VERVAIN or VERBENA

Verbena hastata L.

Perennial; stem somewhat woody below, erect, angled, 1 1/2 to 5 feet tall, entire plant rough-hairy; leaves short-stalked, 1 1/2 to 6 inches long, broadest near base of the blade, long-tapering to the apex, coarsely toothed; flowers dark blue, small, in clustered, slender spikes; nutlets 1/12 inch long, rounded on the back, roughened at the apex.

This species occurs rather commonly in river-bottom land, sometimes becoming troublesome in low pastures or in crops on overflow fields.

PROSTRATE VERVAIN or WILD VERBENA

Verbena bracteata Lag. & Rodr.

Annual or often becoming perennial; branches 1/2 to 1 1/2 feet long, spreading widely, some prostrate, entire plant rough-hairy; leaves deeply lobed and toothed; flowers pale blue, nearly hidden by the long conspicuous bracts, borne in long dense spikes; nutlets brown to gray, 1/16 inch long or less, ridged, roughened.

Prostrate Vervain occurs principally east of the Cascades, but is found also along the lower Columbia River to which area the seeds may have been carried from east of the gorge. It is a weed of pastures and grain fields, also sometimes of lawns.

Nightshade Family: SOLANACEAE

This family contains the common Potato, Tomato, Groundcherry, Eggplant, the garden Pepper, Tobacco, and various medicinal and ornamental plants including Belladonna and Petunia. Several species, most of them introduced, have become troublesome weeds and some are poisonous.

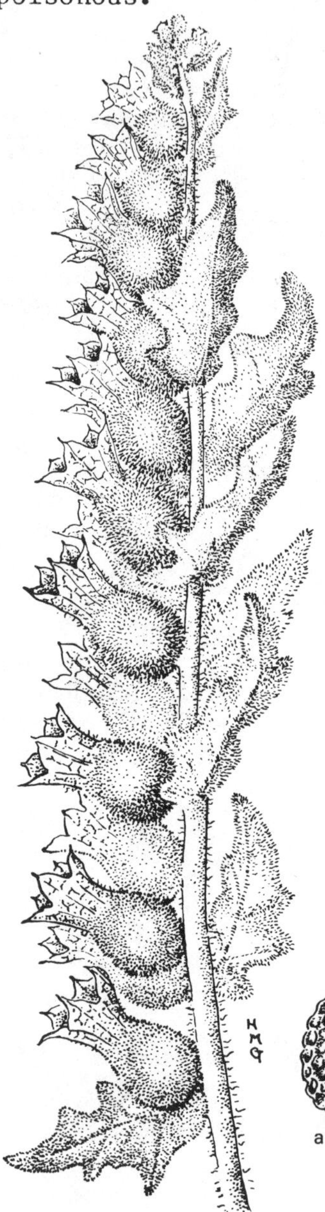

BLACK HENBANE

Hyoscyamus niger L.

Annual or more frequently biennial; coarse, strongly-scented, sticky-hairy, 1 to 3 feet or more tall; leaves 2 to 8 inches long, rather narrow, irregularly lobed or divided, stalkless, the upper clasping the stem; flowers borne in leaf axils and in terminal spikes; calyx bell-shaped, enlarging as the fruit matures, reaching 3/4 to 1 inch long; petals fused into a tube with 5 spreading lobes, greenish yellow with dark purple veins; fruit 1/4 to 1/2 inch long, enclosed in the swollen calyx; seed brown, about 1/16 inch long, pitted.

This poisonous European weed is widely distributed in North America, and has now become well established in the Northwestern States. From every point of view, it is an undesirable introduction and its further spread should be prevented. The fruits split horizontally to free the numerous seeds which scatter widely and germinate freely.

a

BLACK HENBANE
Hyoscyamus niger x 2/3
a. seed x 10

HORSENETTLE

Solanum carolinense L.

Perennial with creeping rootstocks; stem erect, generally much branched, 1 to 4 1/2 feet tall; entire plant clothed with minute star-shaped hairs and long, yellow stiff spines; leaves broad or somewhat narrow, with coarsely wavy margins, both leaf surfaces rough to the touch; flower 1 to 1 1/2 inch in diameter, pale lavender to bluish, borne in clusters; berries yellow, smooth, 1/2 to 3/4 inch in diameter.

This is a species of Nightshade which has become established in a few areas, having reached us perhaps indirectly from the southeastern United States where it is native. Its spiny vegetation makes it useless on the range to any animals except sheep which apparently feed upon its berries without harm.

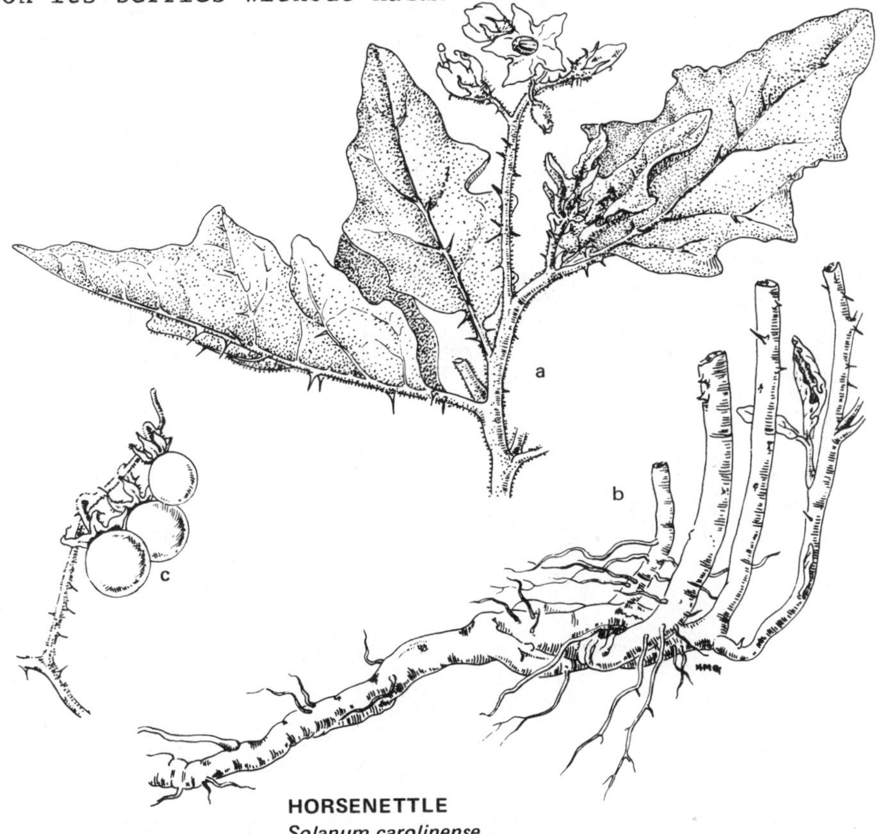

HORSENETTLE
Solanum carolinense
a. upper portion of plant x 2/3
b. lower portion of plant x 2/3
c. fruits x 1/2

BLACK NIGHTSHADE
Solanum nigrum
a. plant with fruits x 3/4

HAIRY NIGHTSHADE
Solanum sarrachoides
b. fruit x 3/4

CUTLEAF NIGHTSHADE
Solanum triflorum
c. upper portion of plant with fruit x 3/4

BLACK NIGHTSHADE

Solanum nigrum L.

Annual, 1/2 to 2 feet tall, much branched; leaves dark green, smooth or somewhat hairy, the blades broadest at or near the base and narrowed to a slender stalk, margins of the blades bluntly toothed to nearly entire; flowers nodding, white or faintly purplish tinged, borne in loose umbrella-like clusters; berry round, black at maturity, with minute reflexed sepals at the base.

This is a highly variable species which is sometimes split into several, including *Solanum nodiflorum* Jacq.

The berries are reportedly poisonous when green, becoming edible at maturity, at least in some forms, as in the large-fruited horticultural forms called "Wonderberry."

This is an early spring weed in gardens and can be found in fields, on roadsides, and in other disturbed or cultivated areas, usually in moist soil.

HAIRY NIGHTSHADE

Solanum sarrachoides Sendt.

Similar to *Solanum nigrum* but with green or yellowish-brown berries which are half-enclosed by the calyx; stems and calyx with sticky hairs.

Native of South America and widely introduced in our area as a weed in gardens and fields, on roadsides and in other disturbed sites. The fruits of this species are also toxic, at least when young.

CUTLEAF NIGHTSHADE

Solanum triflorum Nutt.

Ill-scented annual, much branched from the base; nearly smooth to moderately hairy; leaf blades deeply cleft; flowers white, nodding, arranged in umbrella-like clusters; fruit round, greenish.

Apparently native east of the Cascades to the Great Plains and now found as a weed in cultivated areas. This species is particularly troublesome when it occurs with peas.

Solanum rostratum Dunal.

Annual; stems, leaf stalks, veins, and outer flower part (calyx) covered by stiff sharp yellow prickles; leaves 1 1/2 to 6 inches long, deeply lobed, densely covered by minute yellow branched hairs; berry enclosed in the enlarged densely prickly calyx or "bur."

This species of *Solanum* has thus far been reported mainly from western counties, but more frequent reports in recent years indicate that it is spreading over wider territory. The plant is a native of the midwestern plains. Its spiny exterior can certainly offer little attraction to stock, but cases of mechanical injury from contact with it are known and its presence on the range is undesirable.

BUFFALOBUR
Solanum rostratum
a. flower x 2 1/2
b. "bur" formed by berry enclosed in the prickly calyx x 1 1/2
c. leaf x 2/3
d. tip of leaf showing branched hairs x 2

BITTER OR BITTERSWEET NIGHTSHADE

Solanum dulcamara L.

Perennial; a woody twiner or trailer; stems generally smooth, 2 to 10 or more feet long; leaves slender-stalked, 1 to 4 inches long, broader at the base, narrowed at the apex, often with one or several leaflets or lobes at the base; flowers blue or purplish, stalked, in open clusters; berry scarlet at maturity, ball-shaped or generally somewhat elongated, 1/4 to 1/2 inch long.

A native of Europe but common in parts of the United States, Bittersweet Nightshade is now established sparingly on both sides of the Cascades. The plant is poisonous, though apparently not violently so unless taken in quantity. Its fruits are attractive to children and cases of severe illness have been reported.

The plant twines about other vegetation or, when this is not available, its weak stems trail upon the ground. When under cultivation as an ornamental it often appears shrub-like.

BITTER NIGHTSHADE
Solanum dulcamara x 2/3
a. fruit x 2/3

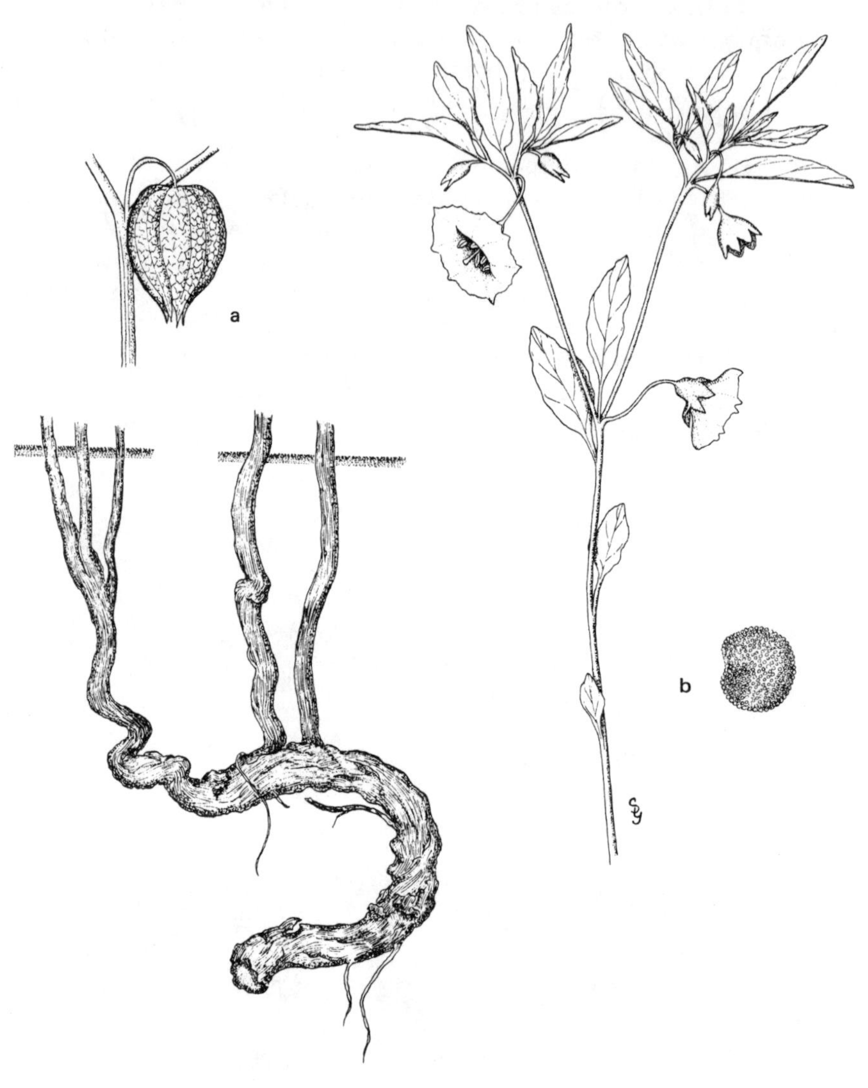

SMOOTH GROUNDCHERRY
Physalis longifolia x 2/3
a. fruit x 2/3
b. seed x 5

SMOOTH or PERENNIAL GROUNDCHERRY

Physalis longifolia Nutt. var. *subglabrata* (Mack. & Bush) Cronq.
(=*Physalis subglabrata* Mack. & Bush)

Perennial; stem 1 to 2 feet tall, much branched and often more or less spreading; leaves 2 to 5 inches long, entire or wavy-margined, ovate, the base of the blade often asymmetrical; flowers long-stalked, nodding or spreading at right angles, widely bell-shaped, 3/4 inch in diameter, yellow to yellowish-green with a purplish or brownish center; berry yellow, enclosed in a nodding strongly-veined, papery sac (the enlarged calyx).

Longleaf Groundcherry (*Physalis longifolia* Nutt. var. *longifolia*) also occurs within our range and has narrower leaves, the blades more tapered at the base.

Both varieties of this species are occasional as weeds in cultivated field, east of the Cascades. The deep underground system gives the plant both food storage and a means of vegetative propagation, thus making it difficult to control.

An annual species, *Physalis pubescens* L. (Downy Groundcherry) is occasional as a weed west of the Cascades where it has been introduced from the eastern United States.

JIMSON WEED

Datura stramonium L.

Annual; 3/4 to 5 feet tall (or under extreme conditions, sometimes flowering and fruiting when only a few inches high), erect, stout, ill-smelling; leaves alternate, wavy-margined or lobed, narrowed to slender stalks, flower white, trumpet-shaped, 3 1/2 to 5 inches long, showy; fruit about 2 inches long, at first fleshy, becoming hard, covered by prickles; seeds numerous, flattened, dark, pitted.

This species, together with variety *tatula* which is generally more slender, with purplish stems and flowers and with less variable prickles, is occasionally found in the Northwest. It is common on river-bottom lands, particularly on sand and gravel bars; and since it is known to be poisonous both to man and animals, it should be regarded as an undesirable introduction.

The large trumpet-shaped flower, prickly pod with a collar-like base, and the rank odor of the plant serve to identify Jimson weed and its close allies, the thornapples. The fruit is at first fleshy but later becomes dry and eventually splits to allow escape of the seeds.

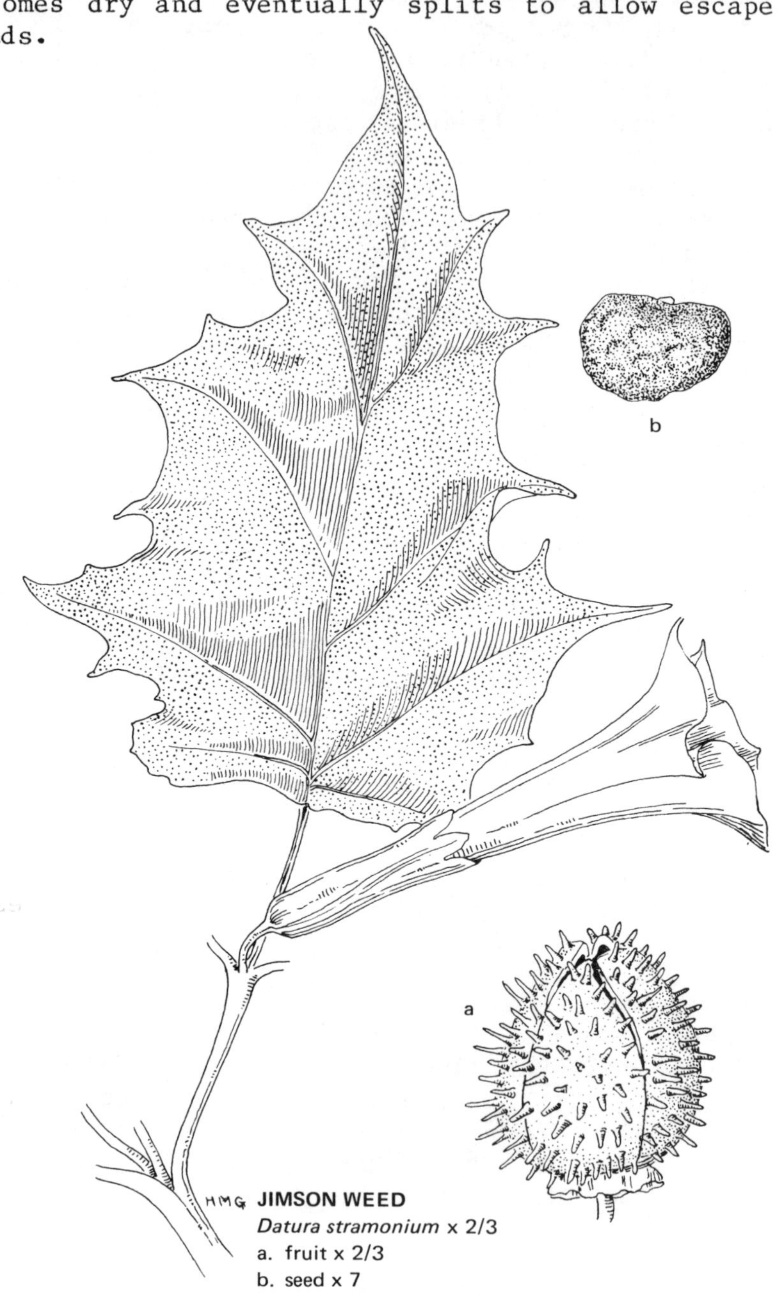

JIMSON WEED
Datura stramonium x 2/3
a. fruit x 2/3
b. seed x 7

Martynia Family: MARTYNIACEAE

A small family represented in our area by a single weedy introduction. The family has slightly irregular flowers and the fruit splits at maturity to form a 2-horned structure.

DEVILSCLAW or UNICORN PLANT

Proboscidea louisianica (Mill.) Thell.

Annual; stems prostrate or spreading, up to 3 1/2 feet long; leaves paired below, sometimes becoming alternate above, sticky and soft hairy, more or less rounded, 2 to 10 inches broad, long stalked; flowers 1 1/4 to 2 inches long, purplish, white or yellow and yellow and purple blotched; fruit fleshy and sticky when young, with a single long beak (hence the name Unicorn Plant), the outer covering of the fruit falling away at maturity to reveal a sculptured woody structure splitting to release the seeds and forming 2 curved horns, body of the fruit 1 1/2 to 4 inches long, the horns as long as the body to 3 times as long; seeds numerous, variously roughened.

This species is a native of the Mississippi Valley and is well established in California; in our area it is occasionally reported from moist areas along lakes or streams and waste places, or as a weed in gardens; all reports are from west of the Cascades and most frequently from southwestern Oregon.

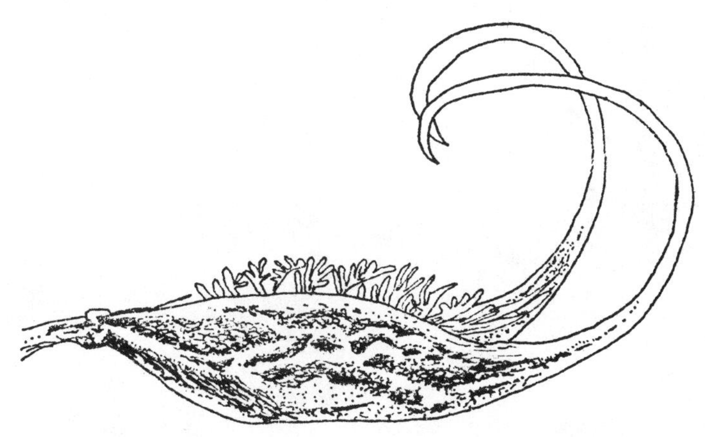

x 2/3

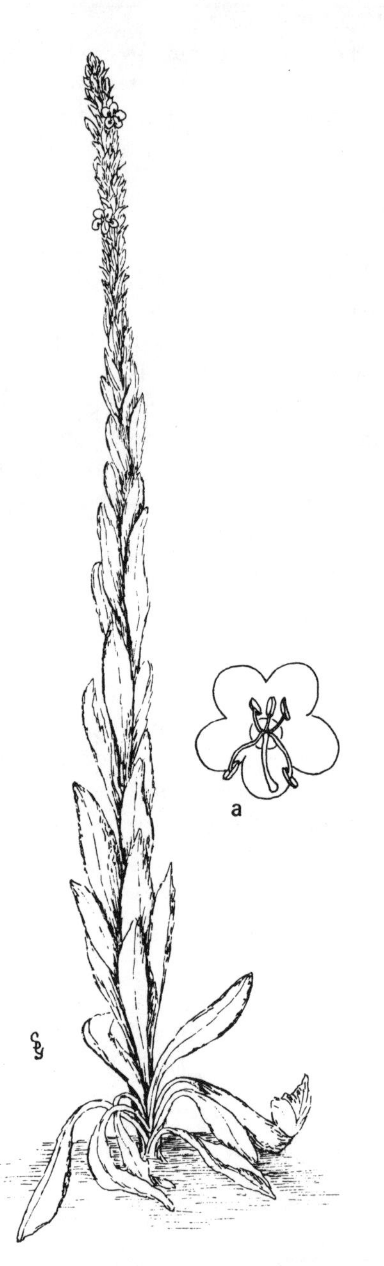

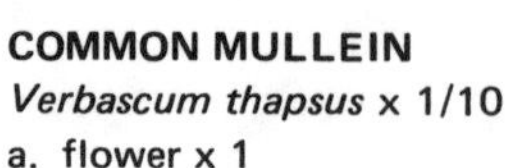

COMMON MULLEIN
Verbascum thapsus x 1/10
a. flower x 1

MOTH MULLEIN
Verbascum blattaria
b. upper portion of plant x 1/2
c. section of stem x 1/2

Figwort Family: SCROPHULARIACEAE

Wide variation exists in this family as shown by the members here described. In general, however, the flowers are more or less irregular, somewhat as in the Mint Family but the fruits generally bear many seeds instead of 4 and split at maturity to allow their escape.

Many representatives of the family including Snapdragon, Penstemon, Speedwell, and Monkeyflower are cultivated for their beauty. Of these and others introduced incidentally in impure seed, contaminated wool, or other means, some have "run wild" and become noxious weeds.

COMMON MULLEIN

Verbascum thapsus L.

Annual, or if biennial, the second-year stem arising from a thick rosette of leaves formed the first year; stem stout, 3 to 7 feet tall, very leafy, leaves densely woolly, gradually decreasing in size from lower part of the stem upward, the lower leaves ranging from 1/2 to 1 foot in length; flowers stalkless, yellow, borne in long dense terminal branched or unbranched woolly spikes; fruit splitting at maturity; seed brown, grooved, and pitted.

A native of Europe, Mullein is now widely naturalized in America. It is found on both sides of the Cascades either as scattered plants or sometimes locally abundant in fields, pastures, on ditch banks, and along roadsides.

MOTH MULLEIN

Verbascum blattaria L.

Generally biennial, the flowering stem arising from a leaf rosette of the previous year; leaves of the rosette dark green, often reddish tinged, deeply lobed, more or less prostrate on the ground; flowering stem 1 1/2 to 4 feet tall, rather slender, erect, its leaves merely toothed; flowers bright yellow or sometimes white, often reddish tinged on the margins, flat, round, slightly irregular in shape.

This species is of European origin but has long been a well-known weed of American fields and roadsides. Except in structure of the flower, this and Common Mullein have little in common other than erect habit. The smooth stem and leaves of Moth Mullein contrast strongly with the felt-like covering of Common Mullein.

IVYLEAF SPEEDWELL
Veronica hederaefolia x 2/3
a. fruit enclosed in calyx x 2 1/2
b. seeds x 6

PURSLANE SPEEDWELL
Veronica peregrina
c. upper portion of stem x 1

IVYLEAF SPEEDWELL

Veronica hederaefolia L.

Annual; stems branching from the base, prostrate or weakly erect, 2 to 15 inches long, leafy; leaves thick, the lower leaves broader than long, coarsely 3- to 5- (rarely 7-) lobed, with an equal number of conspicuous veins from the base; flowers small, blue, short-stalked, borne in the leaf axils; mature fruit approximately 1/4 inch wide, swollen, smooth, bearing 4 seeds; seed 1/8 inch long, curved, hollowed on the concave side, ridged on the convex side, tan to brownish when mature.

This common European species is not yet well known in the Far West, though it has become widely naturalized in parts of eastern North America. It has been reported from scattered localities, both east and west of the Cascades, where it is tending to become a weed in gardens, lawns and grain fields.

Where the plant grows abundantly, its seeds are said to be distributed not only by usual methods of seed dispersal but additionally by ants which drag them to their burrows in order to use as food an oily secretion contained at the bases of the seed stalks.

PURSLANE SPEEDWELL

Veronica peregrina L.

Annual; stems usually much branched, erect or somewhat spreading, 2 1/2 to 12 inches tall; leaves linear, alternate or the lower paired; lower leaves short-stalked, upper leaves sessile; flowers minute, white, borne in the leaf axils; fruit heart-shaped; seeds minute, numerous, light brown.

This native species is commonly represented in our region by var. *xalapensis* (H. B. K.) St. John & Warren. This variety is glandular-hairy, while the typical variety is smooth. Purslane Speedwell usually occurs in wet to moist habitats.

CREEPING SPEEDWELL

Veronica filiformis Sm.

Annual or sometimes perennial; stems slender, thread-like, creeping over the ground and rooting at the nodes; leaves all paired, distant, stalked, the blades broader than long or nearly round, the base heart-shaped, the margins toothed, veins several from the base; flowers 3/8 inch wide, deep blue with dark markings, the slender flower stalk much longer than the leaves and borne in their axils.

This native of Asia Minor is used in rock gardens where its abundant flowers add brilliant blue patches of color in the early morning, the flowers closing later in the day. The slender stems, however, creeping over the ground and striking root at the nodes, form every-widening mats which eventually may take possession, not only of the rock garden but of the surrounding lawn as well. This species somewhat resembles Birdseye Speedwell, with its intense blue flowers on slender stalks. But the stems are generally more delicate, the leaves all paired, the blades broader than long, and the plant more persistent and difficult to control.

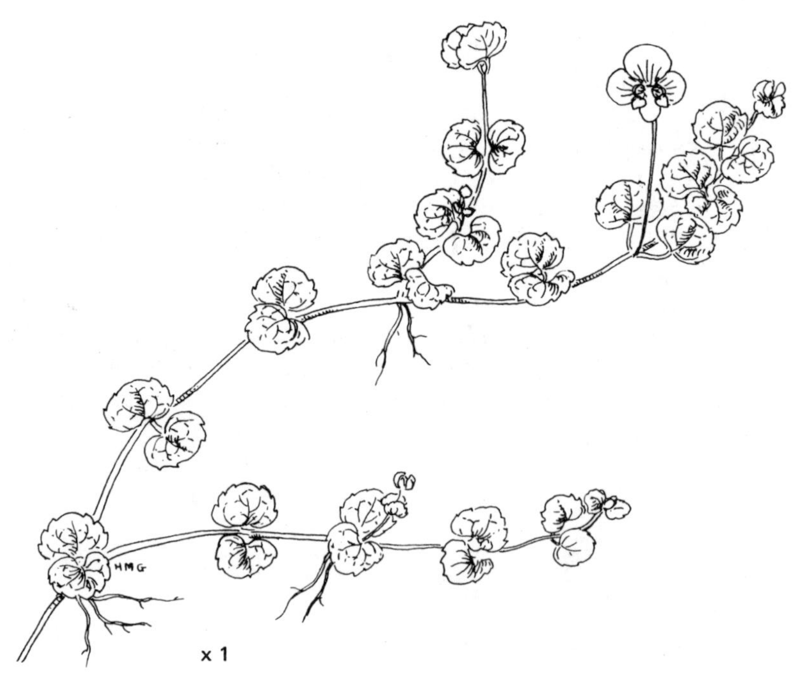

BIRDSEYE OR WINTER SPEEDWELL

Veronica persica Poir.

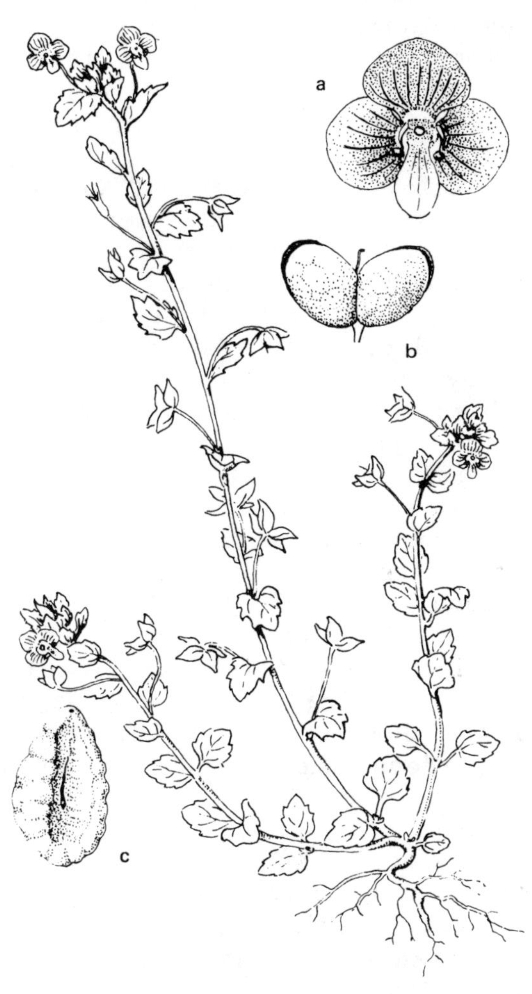

BIRDSEYE SPEEDWELL
Veronica persica x 2/3
a. flower x 3
b. fruit x 2
c. seed x 15

Annual or winter annual; stems weak, spreading over the surface of the ground, the tips ascending; lower leaves in pairs on the stem, the upper leaves alternate, the blades somewhat longer than broad, ovate, rounded at the apex, coarsely toothed, the stalks short; flowers sky blue with dark stripes and white centers, borne on long slender stalks in the leaf axils; fruits heart-shaped, hairy.

A native of Eurasia, this plant probably was introduced as a border or rock garden species. It is one of the most spectacular of ground covers, providing a clear blue shade which is rarely found in flowers. Nevertheless where not wanted, as in lawns, it becomes troublesome and difficult to control.

COMMON FOXGLOVE

Digitalis purpurea L.

Biennial; minutely hairy; stem stout, erect, 2 to 12 feet tall; leaves oblong, toothed, 1/3 to 1 1/2 feet long; flowers 1 1/2 to 2 inches long, nodding, white or reddish purple, paler at the base, dark-dotted within.

Foxglove is a native of Europe, but as far back as the memory of most persons living today can extend, it has abundantly clothed burns and logged-off lands of the American Northwest. As in the case of many plant introductions, it undoubtedly was purposely brought in by early settlers, either as an ornamental garden plant or for its medicinal properties.

As an official drug plant it was collected in enormous quantity in Oregon and Washington during the first World War, the larger proportion of official supplies of *Digitalis* coming from this area. With the development of commercial drug gardens the demand for this plant in the "wild" state has decreased.

Foxglove is sometimes a nuisance in coastal pastures, its rank growth crowding out the natural forage plants. In addition, it is toxic to livestock.

COMMON FOXGLOVE
Digitalis purpurea x 1/2

Redrawn by Helen M. Gilkey, with permission from the Macmillan Company, from Fig. 1266 (p. 1010) in the Standard Cyclopedia of Horticulture by Liberty Hyde Bailey, copyright 1914.

SHARPPOINT FLUVELLIN

Kickxia elatine (L.) Dumort.

Soft hairy annual; stems prostrate or the ends of the branches ascending, stems 4 to 12 inches long; lowest leaves opposite, often toothed, stem leaves alternate, with a pair of basal lobes or the uppermost leaves arrowhead shaped, 1/4 to 1 1/2 inch long; flowers borne in the leaf axils on long thread-like stalks; flowers two-lipped, spurred, pale yellow with a purple upper lip; fruit nearly round, opening by two pores to release the seeds.

Native of Europe; occasional in moist sandy soil west of the Cascades; Oregon to northern California.

Kickxia spuria (L.) Dumort. (ROUNDLEAF FLUVELLIN) is similar to the above species but with oval or rounded leaves 1 to 1 3/4 inches long. It too is a European native occurring in drier habitats, particularly gravel bars and along roadsides. It is less common than *Kickxia elatine*.

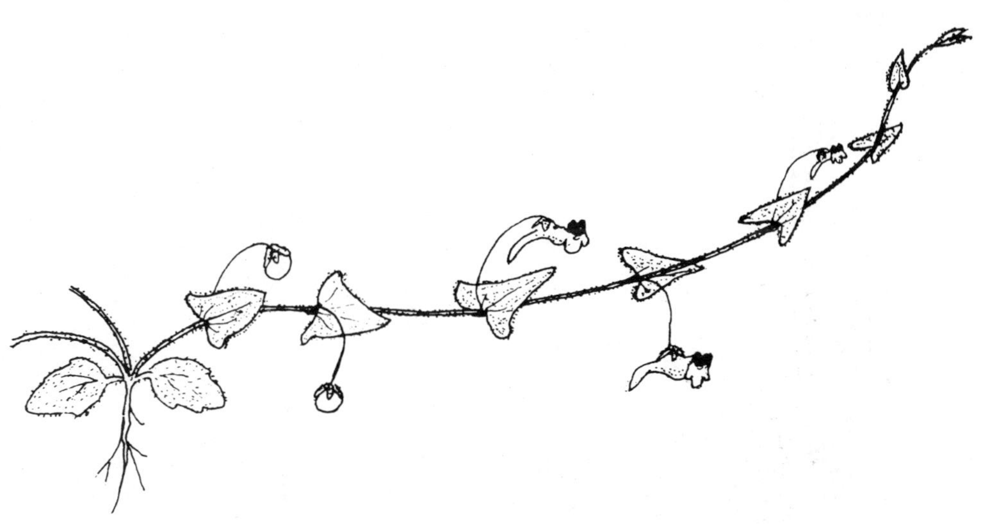

SHARPPOINT FLUVELLIN
Kickxia elatine x 1 1/2

EYEBRIGHT or YELLOW PARENTUCELLIA

Parentucellia viscosa (L.) Car.

Annual; stem erect, 1/4 to 2 feet tall, leafy, stiff-hairy and somewhat sticky-glandular; leaves principally paired on the stem, broadest at the base, narrowed at the apex, sharply toothed, conspicuously veined, flowers bright yellow, 1/2 to 3/4 inch long, borne in spikes.

A native of the Mediterranean region, this plant is found west of the Cascades, sometimes forming large colonies in cultivated fields and pastures. It reproduces only by seed, but quantities of viable seed are produced by each plant, insuring perpetuation and extension of range.

LESSER SNAPDRAGON

Antirrhinum orontium L.

Annual; stems much branched, 8 to 20 inches tall; leaves linear, 3/4 to 2 inches long; flowers axillary, pink-purple, about 5/8 of an inch long; fruit asymmetrical, opening by pores to release the numerous seeds.

Only occasional in the Willamette Valley of Western Oregon.

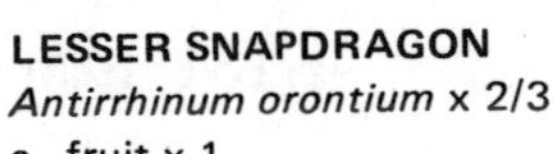

LESSER SNAPDRAGON
Antirrhinum orontium x 2/3
a. fruit x 1

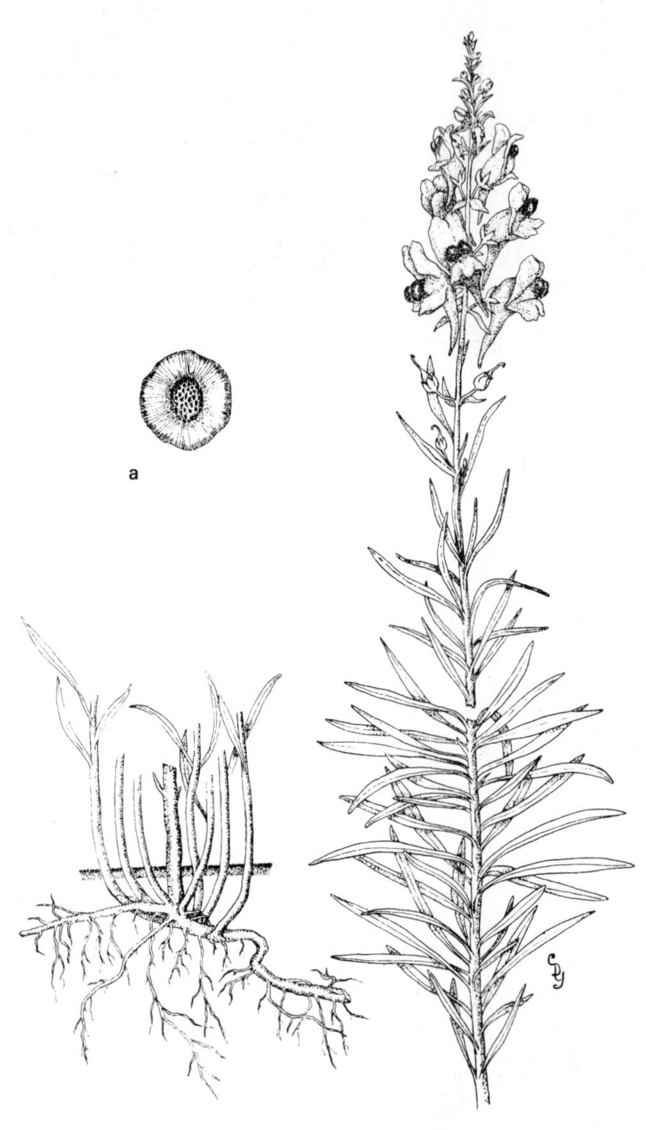

YELLOW TOADFLAX
Linaria vulgaris x 1/2
a. seed x 10

YELLOW TOADFLAX or BUTTER-AND-EGGS

Linaria vulgaris Hill

Perennial from a woody underground base capable of sending up numerous new plants; stem erect, 1 to 2 feet tall, little branched or sometimes branching; leaves numerous, pale green, narrow, pointed at both ends; flowers 1 inch or more long, spurred, yellow with orange center, produced in an elongating raceme; fruit 1/4 inch broad, containing 2 compartments and many seeds; seed 1/12 inch in diameter, flat, with papery circular wing.

As in the case of many other noxious weeds, Yellow Toadflax is a European plant introduced into America as a garden ornamental. Very early, however, it escaped from cultivation and became a pest in gardens, pastures, and fields. It has been known in the Pacific Northwest for many years, and although local areas covered by it are usually still not extensive, the plant is now widely distributed over the area.

Its showy and attractive Snapdragon-like flowers are borne at first in dense clusters which gradually elongate, new flowers being formed above as fruits mature below. The single slender spur of each flower contains nectar which normally furnishes food for insects with long sucking mouth parts and which, as they feed, assist in pollinating the flower. Occasionally, however, bees with short mouth parts that are incapable of reaching the nectar by legitimate means may be seen cutting circular holes in the spurs to extract it. Since they do not enter the flower or come in contact with the pollen, they are of no assistance in pollination but perhaps are of indirect service to man in thus forestalling normal seed development.

Since the plant has ability to spread underground, its local distribution is not dependent entirely upon seeds. Those seeds which are formed have effective means of dissemination in the membranous wings which enable them to be carried readily by air currents.

DALMATIAN TOADFLAX
Linaria dalmatica x 4/5
a. two views of the flower x 4/5
b. seed x 10

DALMATIAN TOADFLAX

Linaria dalmatica (L.) Mill.
(=*Linaria macedonica* Griseb.)

Perennial from a woody branching base; stem reaching 2 1/2 to 3 or more feet in height, densely leafy; lower leaves somewhat narrowed at the base, those of the middle and upper parts of the stem conspicuously broad-based, becoming shorter upward; flowers borne in the axils of upper leaves and bracts, flower 3/4 to 1 1/2 inches long (including the spur), deep yellow with an orange throat, spur long, straight or somewhat curved, gradually narrowing to a point.

A southern European species, Dalmatian Toadflax was apparently introduced for ornamental purposes, but is now spreading to roadsides and grain fields. Its underground structures for perpetuation and distribution appears formidable.

BROAD-LEAVED TOADFLAX

Linaria genistifolia (L.) Mill.

Somewhat resembling Dalmatian Toadflax in size and woody base, but the leaves proportionately narrower, and the flowers much smaller (1/2 inch or slightly longer, including the spur), pale yellow in color with a darker throat. The seeds in both species are angled, winged, and netted, about 1/16 inch in diameter.

This, too, is a native of Europe and is a rapidly spreading grain field weed.

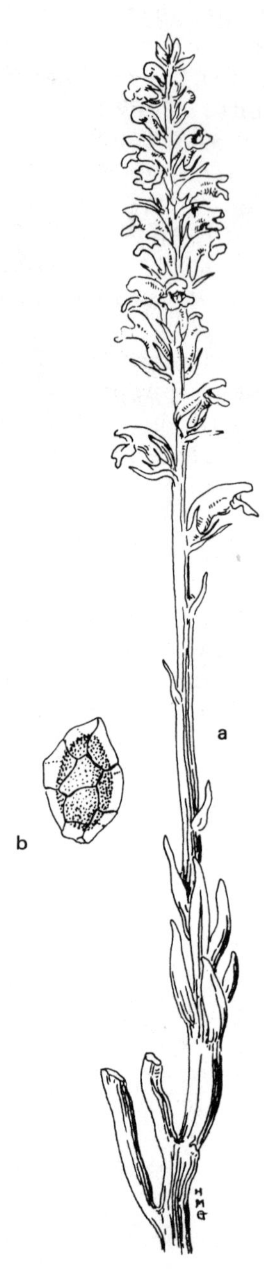

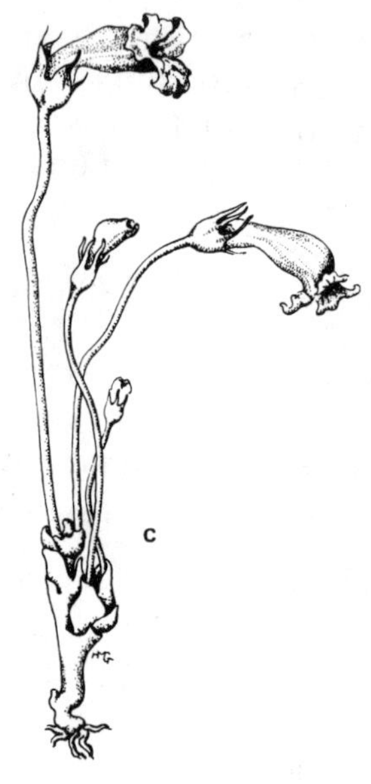

CLOVER BROOMRAPE
Orobanche minor
a. plant x 2/3
b. seed x 55

c. **NAKED BROOMRAPE**
Orobanche uniflora x 2/3

Broomrape Family: OROBANCHACEAE

This is a family of fleshy parasitic herbs without green color. The flowers are irregular and borne singly or more often in a spike. The fruit contains a single cavity, with 2 or 4 lines of seeds, the latter generally numerous. These plants secure their food by attaching themselves to the roots of green plants. Certain species attack vegetables, such as carrots, stunting and deforming their roots, and producing an abundance of seed for a new crop of the parasite.

CLOVER BROOMRAPE

Orobanche minor J. E. Smith

A yellowish-brown parasite, growing on the roots of clover and other plants; stem slender to rather stout, 1/3 to 1 3/4 feet tall, simple or branched, somewhat sticky-hairy, bearing scale-like leaves at the base, and at the apex a more or less compact spike of flowers; flowers yellowish and slightly bluish-tinged; fruit swollen; seeds numerous, approximately 1/80 inch long, with a microscopic netted covering.

This species is a native of the Mediterranean region and has been introduced throughout the temperate regions of the world; in our area it is thus far confined to southwestern Washington and northwestern Oregon.

NAKED or SINGLE-FLOWERED BROOMRAPE

Orobanche uniflora L.

Stems short, sending up one to several slender flowering stalks, each bearing a single flower 1/2 to 1 inch long; flowers somewhat snapdragon-shaped, and white, or lavender to purple.

Naked Broomrape is a native American species and shows little preference in hosts, growing upon the roots of apparently any plant which is conveniently near at the time of germination of its seed. Its importance as a weed in cultivated fields is not thoroughly known. It is frequently overlooked because of its small size and the usual purplish tint of the entire plant which is effective protective coloring when viewed against a background of gravel, sand or garden soil.

BROADLEAF PLANTAIN
Plantago major x 2/3
a. seeds x 12

Plantain Family: PLANTAGINACEAE

The only representatives of this family in our area belong to the genus *Plantago*, of which we have several native species. The only species which here are troublesome as weeds, however, have come to us from Europe. All have small often papery greenish, brownish, or whitish flowers in close spikes.

BROADLEAF PLANTAIN

Plantago major L.

Perennial, with a thick basal crown; leaves spreading, often prostrate, blades very broad, sparingly toothed or entire, veiny, abruptly narrowed at the base to a wide stalk; flowering stalks generally longer than the leaves; spikes long, narrow, dense; seeds minutely netted.

Broadleaf Plantain, like Buckhorn Plantain, is a native of the Old World but its distribution is not world-wide. The Indians of North America called this plant the "white man's footsteps," since it appeared almost simultaneously with him in all his settlements.

Broadleaf Plantain is less aggressive than its close relative, Buckhorn Plantain, being more restricted in its growth requirements and therefore of less seriousness as a noxious weed.

BUCKHORN OR ENGLISH PLANTAIN

Plantago lanceolata L.

Perennial, with a thick branching crown; leaves and flower stalks erect; plant nearly smooth except for dense tufts of long brown hairs at the base; leaves 1/4 to 1 or more feet long, narrow, pointed at the apex, tapering at the base to a long slender stalk, blades conspicuously veined, slightly toothed; flowering stems longer than the basal leaves, bearing dense spikes, these short and headlike at first but somewhat elongating as the flowers open; seed about 1/16 inch long, oblong, smooth, shining, brownish, the seedcoat becoming sticky when wet.

Although probably a native of Europe, it is now found on every continent. Since the turn of the century it has spread rapidly in northwestern America and is now one of its most abundant and pernicious weeds in lawns, gardens, fields, and pastures. It spreads not only by seed but also by crown expansion, and is resistant to superficial cutting.

During flowering, the stamens protrude from the flower, and quantities of pollen may be released in the air. Since Plantain pollen is known to be one of the causative agents of hay fever, and since Plantain may be found in flower at any season when conditions are favorable, the abundance of these plants may have certain bearing on summer cases of allergy.

BUCKHORN
Plantago lanceolata x 2/3
a. seed x 8

Madder Family: RUBIACEAE

Aside from the readily recognized weeds belonging here, this family is outstanding from the fact that in it are included Coffee, Gardenia, Quinine, and several species of plants which are sources of valuable dyes.

FIELD MADDER

Sherardia arvensis L.

Annual; roots often dark reddish; stems weakly erect or spreading mat-like, 2 to 18 inches long, angled, minutely barbed; leaves in whorls of 4 to 6, narrow or the lower sometimes broad; flowers pink or lavender in dense heads surrounded at the base by a whorl of leaves fused at their bases; fruit permanently enclosed in the 4- to 6-toothed calyx tube, splitting at maturity into 2 small sections, each tipped by 2 or 3 calyx teeth.

This is a common weed of lawns and meadows. It is especially troublesome in lawns because of its tendency to form large mats in the turf, differing in texture from the surrounding grass. In early spring, or when well watered, it is not unattractive in itself as a ground cover, but the tendency of the lower leaves to turn brown and die as new growth occurs above is revealed in mowing, and the plant becomes weedy and unsightly as the season advances.

Field Madder is a native of Europe but now has spread to all the larger continents.

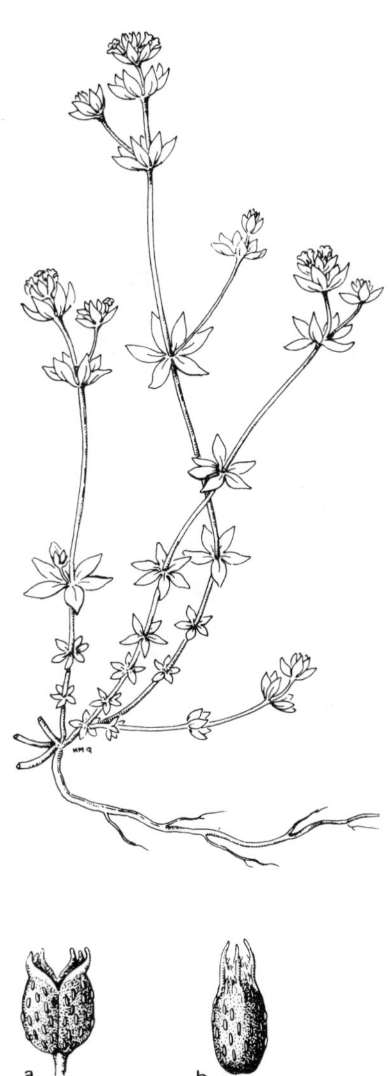

a b

FIELD MADDER
Sherardia arvensis x 2/3
a. fruit before splitting into two segments x 6
b. single segment of fruit x 6

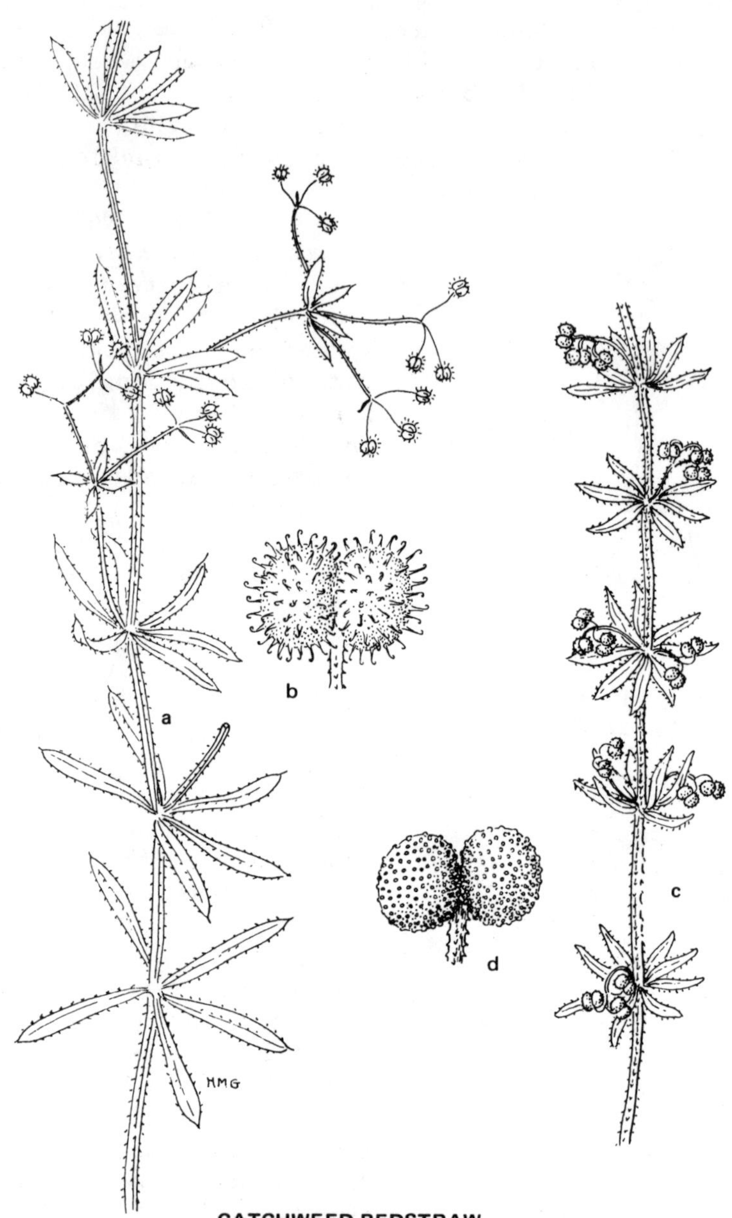

CATCHWEED BEDSTRAW
Galium aparine
a. portion of plant x 1
b. fruits x 4

CORN BEDSTRAW
Galium tricorne
c. portion of plant x 1
d. fruits x 4

CATCHWEED BEDSTRAW OR CLEAVERS

Galium aparine L.

Annual; stems 1 to 5 or 6 feet long, weak and trailing, or climbing over any available support, 4-angled, the angles armed with short backward-turning bristles, branches few; leaves narrow, generally 6 to 8 in each whorl, sometimes stiff-hairy; flowers minute, white, borne on short branches in the leaf axils; fruit densely covered by fine hooked hairs.

This species of Bedstraw is a native of the United States, and though very common in the Northwest, is of little importance as a weed in cultivated crops. Practically its only claim to recognition in a weed manual is the annoyance caused by its small hooked-bristly fruits which are easily dislodged from the plant by passing man or animals and become attached to clothing, wool, and fur. This is a means of distribution for the plant, but its effectiveness is considerably reduced by the fact that comparatively few of the enormous number of seeds produced ever reach satisfactory conditions for germination. The natural habitat of the plant is undisturbed thickets and woods.

In contrast with this native species, as a potential pest, is the introduced species described below.

CORN BEDSTRAW

Galium tricorne L.

Annual; stems somewhat stout, more or less erect, 1/2 to 1 1/2 feet long, 4-angled, the angles thickened, each bearing a row of stiff down-turned barbs; leaves in whorls of generally 5 to 8, very narrow, minutely barbed; flowers minute; fruit dark brown, paired, borne in axillary clusters on down-turned stalks, each approximately 1/8 inch in diameter, roughened by microscopic wart-like projections.

Corn Bedstraw is a native of Europe, but is proving troublesome when found occasionally in western grain fields. Its fruits may form an impurity in seed crops.

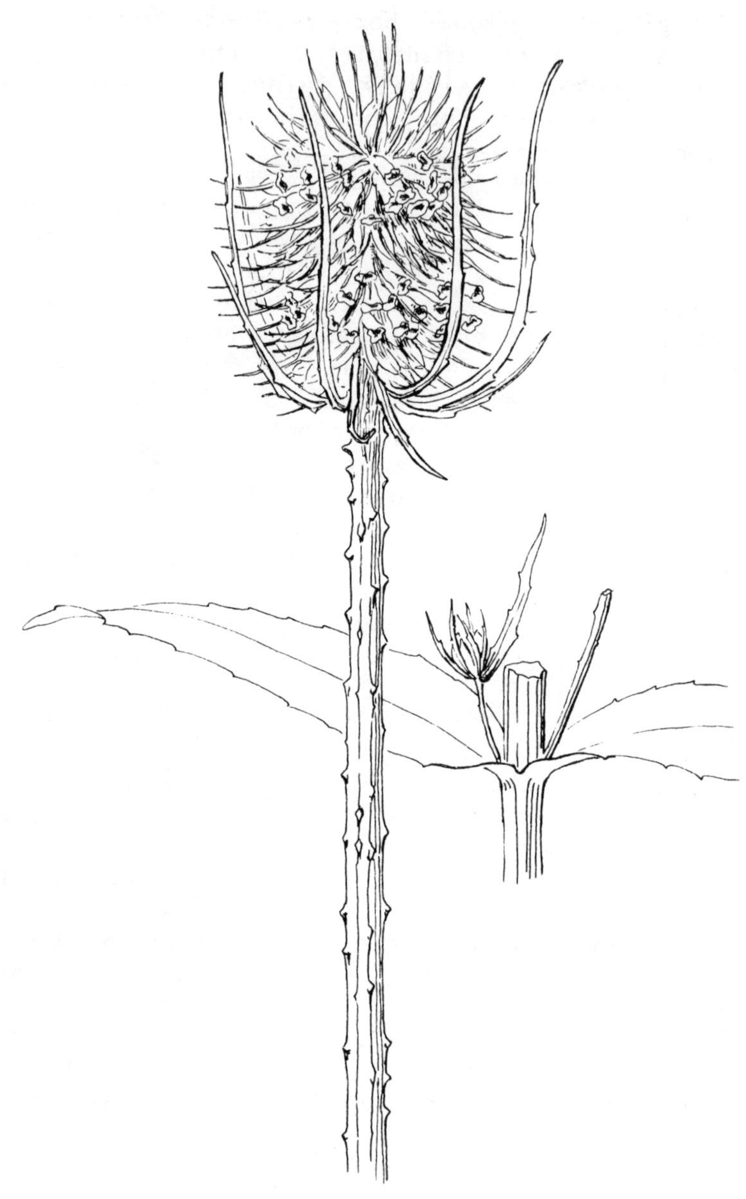

TEASEL
Dipsacus sylvestris x 2/3

Teasel Family: DIPSACACEAE

No member of this family is native to America, but two genera have reached us from Europe. *Scabiosa* is frequently grown in flower borders, and *Dipsacus* (Teasel) is common along many roadsides in the Northwest. The family is apparently closely related to the Composite Family, differing in the constantly paired or whorled leaves, and in a few details of the flower and seed. The flowers are clustered in dense heads or spikes, with a series of modified leaves (bracts) at the base.

TEASEL

Dipsacus sylvestris Huds.

Biennial; stem 3 to 6 feet tall, sharply ridged, the ridges bearing short stiff down-turned, yellowish prickles; leaves large, conspicuously veined, with stiff prickles on at least the midrib and sometimes other veins; leaf blades stalkless, paired, the bases of the upper pairs generally fused; flowers purple, borne in thick stiff spikes, these bristly with spine-like bractlets, and surrounded at the base by stiff curved prickly bracts generally longer than the spike.

This spectacular species is a native of Europe but has become established in the Pacific Northwest, particularly west of the Cascades. Ordinarily it is of no outstanding importance as a weed except as it replaces plants of more value. It is occasionally introduced into gardens through river loam.

Fuller's Teasel (*Dipascus fullonum* L.), resembling common Teasel but with the bractlets of the spike curved strongly downward, was brought to the west coast in the early years for the purpose of carding wool. It has not become widely distributed; but the fact that common Teasel is found on the sites of early carding mills in the Willamette Valley suggests that this species, also, though less efficient, may have been employed for the same purpose.

Valerian Family: VALERIANACEAE

A small family of herbaceous species with opposite leaves, irregular flowers and an inferior ovary.

EUROPEAN CORNSALAD
Valerianella locusta x 2/3
a. flower and fruit x 4

EUROPEAN CORNSALAD

Valerianella locusta (L.) Betcke
(=*Valerianella olitoria* Poll.)

Annual; stems slender, 2 to 16 inches tall, several times forked above; lower leaves stalked, upper leaves sessile, entire or irregularly toothed, 3/8 to 2 1/2 inches long, less than 3/4 inch wide; flowers small, borne in paired clusters, white to light bluish-purple; fruit grooved and with a corky enlargement on the back, 1-seeded.

This European native is well established throughout our area in moist open places, often in disturbed areas and occasionally in cultivated fields.

Gourd Family: CUCURBITACEAE

Members of this family include Melon, Squash, Pumpkin, Gourd, and Cucumber. All are trailing or climbing vines, and all have pollen and seed borne in separate flowers.

WESTERN WILD CUCUMBER or OLD-MAN-IN-THE-GROUND

Marah oreganus (T. & G.) Howell
(=*Echinocystis oregana* Cogn.)

Perennial by an enormous root; stem long, thick, angled, trailing or climbing; leaves stalked, the blades lobed, roughened, sometimes reaching 6 inches or more in length and width; flowers waxy-white, somewhat star-shaped, 5- or 6-lobed, the pollen-bearing flowers borne on a raceme in the leaf axils, the seed-bearing flowers in the same location but borne singly; fruit gourd-like, several inches long, beaked at the apex, fleshy at first and somewhat spiny, at length becoming somewhat papery and breaking irregularly to release the large seeds.

This is a common vine of far western fields and low hillsides, climbing fences and shrubs or, when no support is present, spreading prostrate on the surface of the ground.

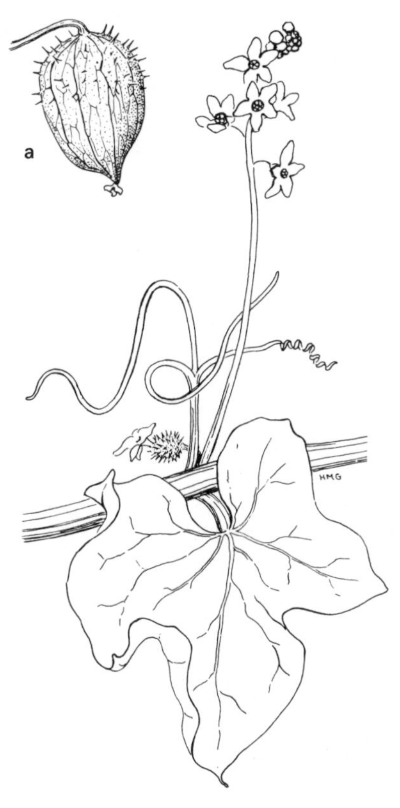

WESTERN WILD CUCUMBER
Marah oreganus x 2/3
a. fruit x 1/2

Bluebell or Harebell Family: CAMPANULACEAE

The "Bluebell of Scotland" and the Canterbury Bell have made this family famous. These and other species are cultivated for their showy flowers, and several are native to the Pacific Coast.

CREEPING BELLFLOWER OR PURPLE-BELL

Campanula rapunculoides L.

Perennial by deep-seated spreading rootstocks; stem often purplish, erect, 2 to 4 feet tall, slightly hairy to smooth, leafy, bearing at the apex a long one-sided spike of purple bell-shaped flowers; leaf blades somewhat rough long-ovate to lance-shaped, tapering at the apex, rounding at the base to a slender stalk, margins of the blade toothed; flowers nodding, long bell-shaped 1 to 1 1/2 inches or more in length; seeds many.

An example of a beautiful ornamental plant "gone wrong" is seen in this species when, as frequently happens, it escapes bounds and encroaches on areas where it is not wanted. Its ability to cover the ground rapidly, in either sun or shade, has made it useful in the garden. However, once established, it not only tends to be permanent, but requires constant control to prevent its taking over the garden completely. Beneath the ground it possesses a slender stem which may extend 6 inches or more below the surface. At this point are clustered fleshy storage roots, together with laterally speading rootstocks which, as they lengthen, send new stems upward and clusters of roots downward. When the plants are hand-pulled, the slender vertical stem breaks, leaving behind in the soil the storage roots and the propagating rootstocks to continue their colonization of the area.

In a border, the new growth of this plant is often obscured at first by its ability to push up through clumps of other plants such as violets, primroses, and Michaelmas daisies, where its veiny and often purplish new growth offers an effective early mimicry of these neighbors.

In addition to propagating underground, the plant bears abundant seed which germinates readily. Still further, under certain conditions unfavorable to growth, such as insufficient moisture in midsummer, thick bulb-like resting buds are sometimes formed beneath the soil, ready to develop quickly into basal branches or new plants as soon as conditions again become favorable.

In certain Eastern States this plant has "run wild" along roadsides and in fields. That apparently has not yet occurred to any dangerous degree in the Northwest, but the plant is being reported as a serious garden pest in various areas.

CREEPING BELLFLOWER
Campanula rapunculoides x 1/3
a. flower x 4/5

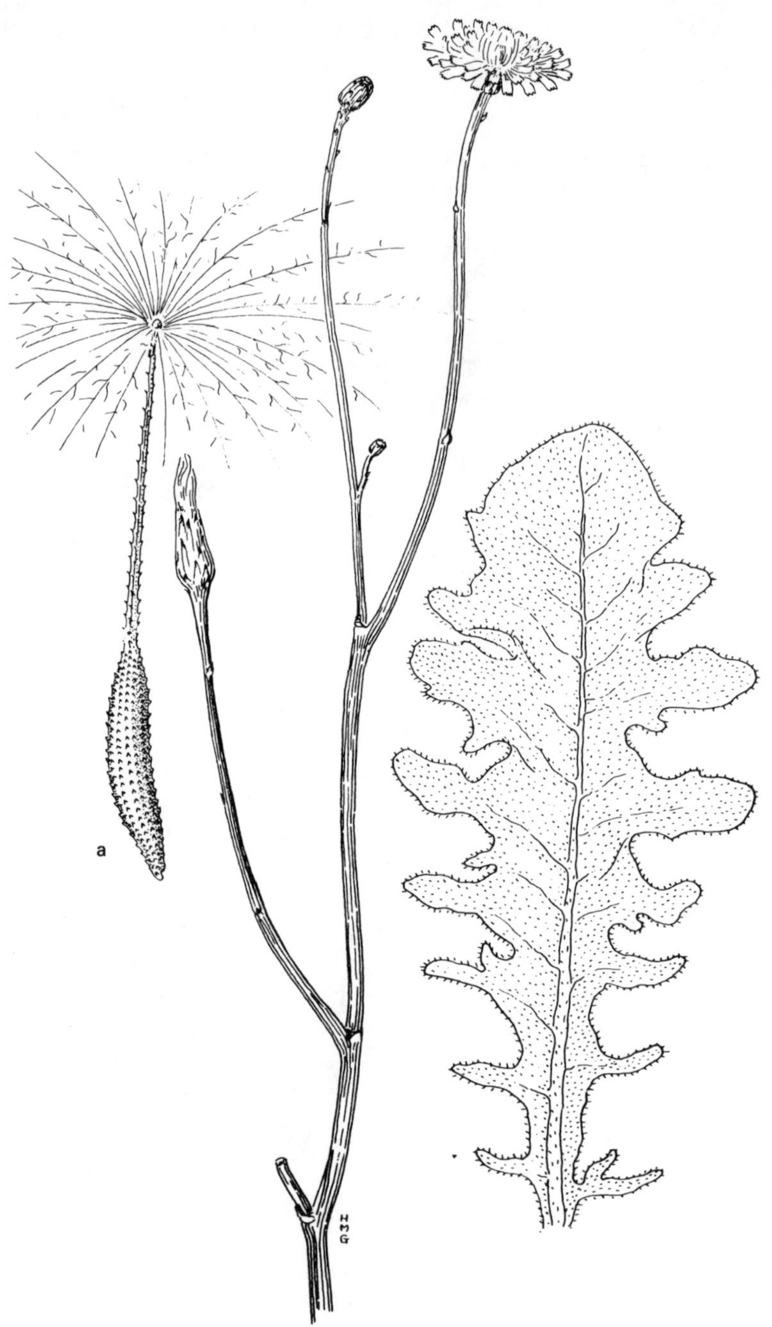

SPOTTED CATSEAR
Hypochaeris radicata x 2/3
a. fruit x 9

Composite or Sunflower Family: COMPOSITAE

This is the largest family of vascular plants. It contains such well known weeds as Thistle, Dandelion, Chicory, Ragweed and Knapweed. Here also belong the garden vegetables Lettuce, Salsify, Artichoke and Endive. A few of the many ornamental species in this family are Aster, Chrysanthemum, Dahlia, Daisy, Marigold and Zinnia.

The concentration of numerous small flowers in a head resembling a single large flower is characteristic. In the English Daisy, for example, the outer flowers resemble petals surrounding numerous smaller central flowers. In Dandelion, Thistle and other members of this type, all the flowers are similar and open successively from the outside to the center of the head. Surrounding each head in members of this family are more or less sepal-like bracts. The fruit in many species bears a tuft of bristles, plumes or scales which aid in seed distribution.

The alternate name for the family is ASTERACEAE.

SPOTTED CATSEAR, FALSE DANDELION or GOSMORE

Hypochaeris radicata L.

Perennial; leaves 2 to 12 inches long, lobed or wavy-margined, rough-hairy, borne in a basal rosette; flowering stems sparsely branching, 3/4 to 2 feet tall; heads 1 to 1 1/2 inches in diameter, yellow; fruit long-beaked, roughened, tipped by a circle of plume-like bristles.

This native of Europe is a weed of lawns, pastures, and waste places, particularly west of the Cascades. In close-clipped lawns, its foliage tends to hug the ground tightly and forms unsightly areas. The ascending flowering stalks resist mowing and many of them spring back after the machine has passed. Where competition is keen or growth conditions are more favorable, as in gardens and fields, the leaves may stand nearly erect.

Another species, Smooth Catsear or Smooth False Dandelion (*Hypochaeris glabra* L.), is likewise common and often occurs with the preceding species. It differs in having smooth instead of hairy leaves and less conspicuous heads. It, too, is a native of Europe.

COMMON DANDELION
Taraxacum officinale x 4/5
a. fruit x 3

COMMON DANDELION

Taraxacum officinale Weber

Generally perennial with leaves and flower stalks arising from a thick crown and fleshy taproot; leaves clustered on the crown, spreading or erect, 2 to 12 inches long, at first somewhat hairy but generally smooth at maturity, variously lobed and toothed, the terminal lobe often much larger than the laterals, the latter decreasing in size toward the base of the leaf, and generally pointed backward; stalks of flower heads unbranched, hollow, with milky juice; head golden yellow, opening flat, 3/4 to 1 1/2 inches in diameter, surrounded at the base by 2 rows of bracts, the outer turning back upon the stem; flower head followed by a ball of "down," consisting of numerous tiny parachutes, each attached to a minute fruit.

Although a native of Europe, Common Dandelion has been known in American nearly as long as the white man and is one of the most widely distributed of all weeds. It is probable that its introduction was not accidental for the plant was early classed as an "herb" providing both food and medicine. The roots and leaves in the fresh state were used as greens and, when dried, provided a tea valued as a spring tonic.

The bright golden flower heads are not unattractive in the first days of spring, even in a lawn. But, the rosette habit of the plant, with its spreading leaves, interrupts the texture of the turf and eventually crowds out the grass. The individual plants spread by crown propagation and distribution is accomplished by an abundant seed crop. At maturity, the tiny parachutes become detached from the head and fall in the vicinity in which they were produced, or are carried far and wide by air currents.

The plants provide good forage in pastures, but in lawns they are out of place.

ROUGH or HAIRY HAWKBIT

Leontodon nudicaulis (L.) Merat

Perennial; leaves from a basal rosette, 1 1/2 to 6 inches long, to 1 inch wide, shallowly lobed, stiff-hairy; stalks of the flower heads unbranched, leafless, 4 to 15 inches tall; young buds nodding; heads solitary; flowers all strap-shaped, yellow, purplish or brownish on the back; fruits 1/8 to 3/16 inch long, without a beak or with a very short beak, tipped by scales and bristles.

Native of Europe and well established west of the Cascades in Washington and Oregon as a weed of lawns and waste places. Often associated with *Hypochaeris*. Rough Hawkbit is particularly abundant along the coast.

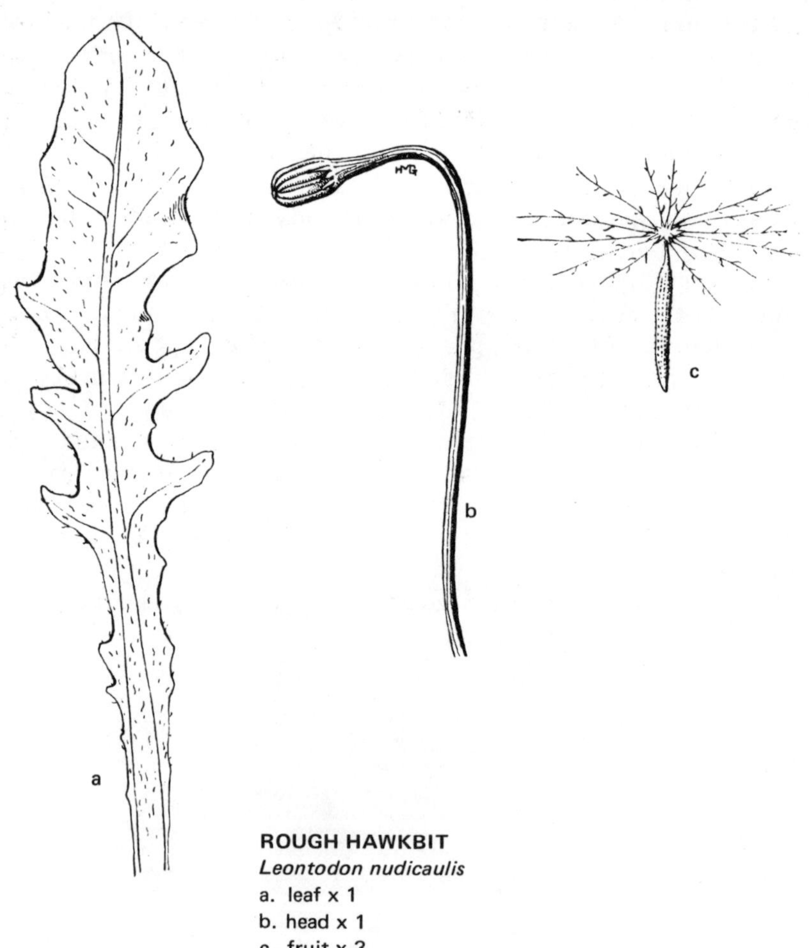

ROUGH HAWKBIT
Leontodon nudicaulis
a. leaf x 1
b. head x 1
c. fruit x 3

SPINY OR PRICKLY SOWTHISTLE

Sonchus asper (L.) Hill

Annual or biennial; stem generally stout, 1 to 5 feet tall; the plant considerably resembling Annual Sowthistle but generally with the terminal lobes of the lower leaves less conspicuous and the prickles stiffer and sharper. The rosettes of leaves, produced the first year are sometimes mistaken for that of Canada Thistle.

This species, like our other sowthistles, came to us from Europe and is now widely distributed.

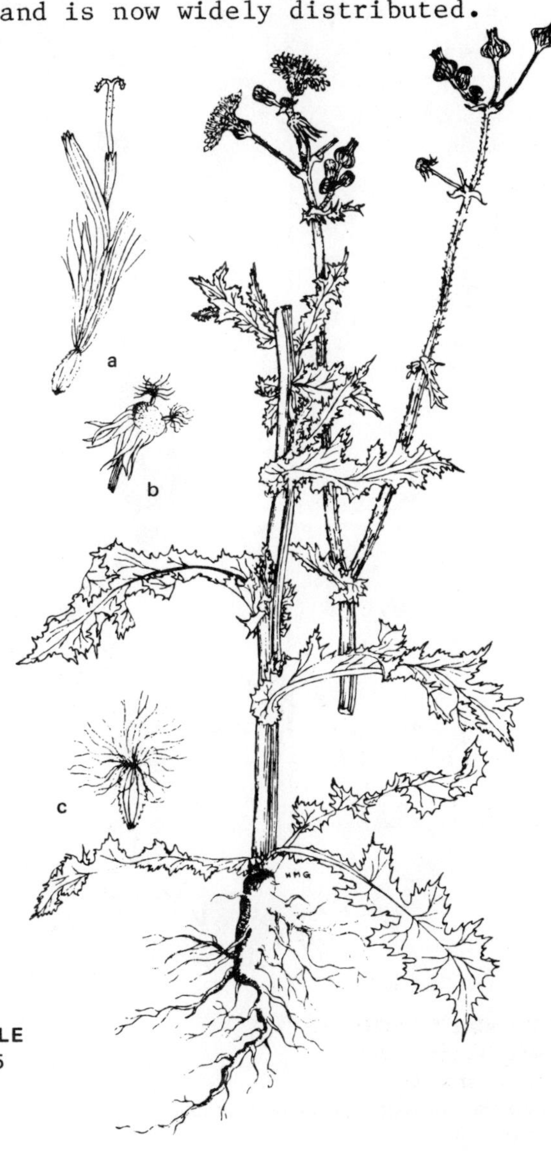

SPINY SOWTHISTLE
Sonchus asper x 2/5
a. floret x 4
b. receptacle x 1/4
c. fruit x 3

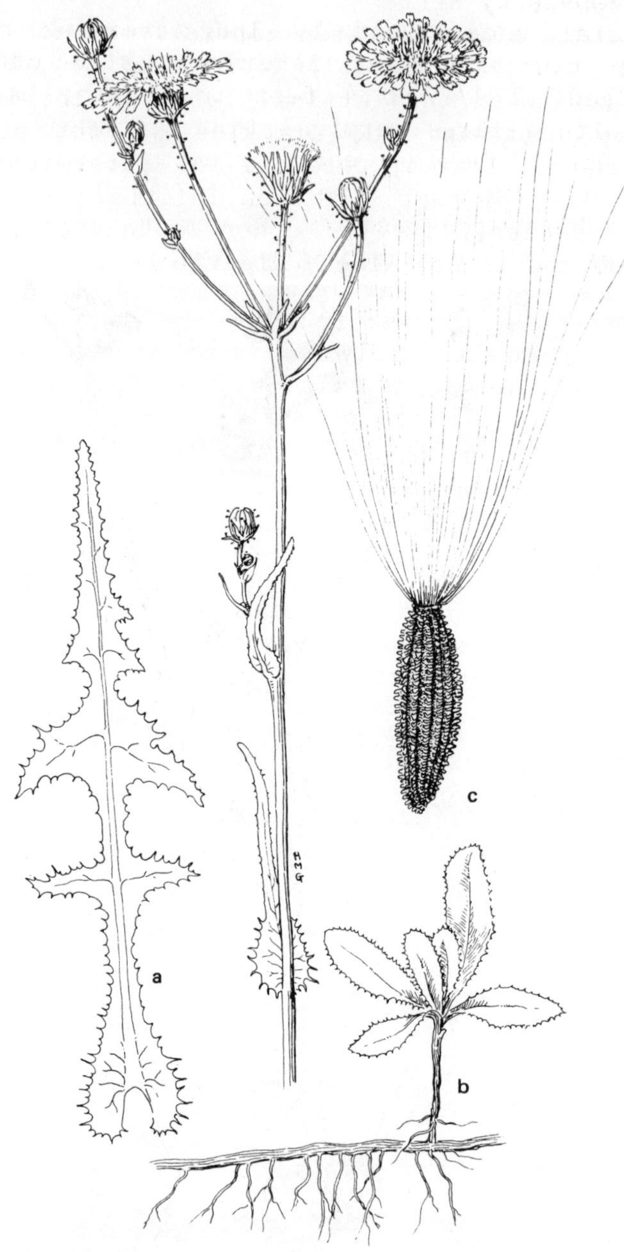

PERENNIAL SOWTHISTLE

Sonchus arvensis x 2/3

a. basal leaf x 2/3

b. new plant developing from underground rootstock x 2/3

c. fruit x 10

PERENNIAL SOWTHISTLE

Sonchus arvensis L.

Perennial underground by long creeping rootstocks, sending up new plants at intervals; stem often stout, finely ridged, 1 1/2 to 4 feet in length; basal leaves generally with a large terminal lobe and several downward-turned lateral lobes on each side, or sometimes merely toothed, the margins of the leaf minutely prickly; stem leaves resembling the basal but becoming smaller upward on the stem, or often consisting only of 2 large basal lobes clasping the stem; flower heads 1 1/4 to 2 inches wide, yellow, sometimes borne in widely branching clusters; fruit ribbed longitudinally and with transverse wrinkles, and bearing at the apex a tuft of long white silky hairs.

In the typical form of this species, the stalks and bases of the flower heads bear knobby glandular hairs, this character serving to distinguish it from other species in our area. However, a smooth variety, *glabrescens*, frequently is the form encountered here.

Perennial Sowthistle is a native of Europe but was probably introduced here through impure commercial seed. Its ability to spread rapidly, both underground by rootstocks and above ground by seed, makes it a serious menace. It occurs rarely in the Willamette Valley but is being reported with increasing frequency from east of the Cascades.

ANNUAL or COMMON SOWTHISTLE

Sonchus oleraceus L.

Annual or biennial; entire plant somewhat bluish green; stem 1 to 4 feet tall, finely ridged; basal leaves generally with a large terminal lobe and several smaller lateral lobes, the upper leaves margined by teeth and soft prickles, and clasping the stem by 2 large basal lobes; heads yellow, several, borne in branched clusters; fruit reddish brown, ribbed longitudinally and roughened horizontally, bearing an apical tuft of long silky white hairs.

This European species is frequently seen about old buildings, and is a common inhabitant of waste places, farmyards, cultivated fields, and gardens.

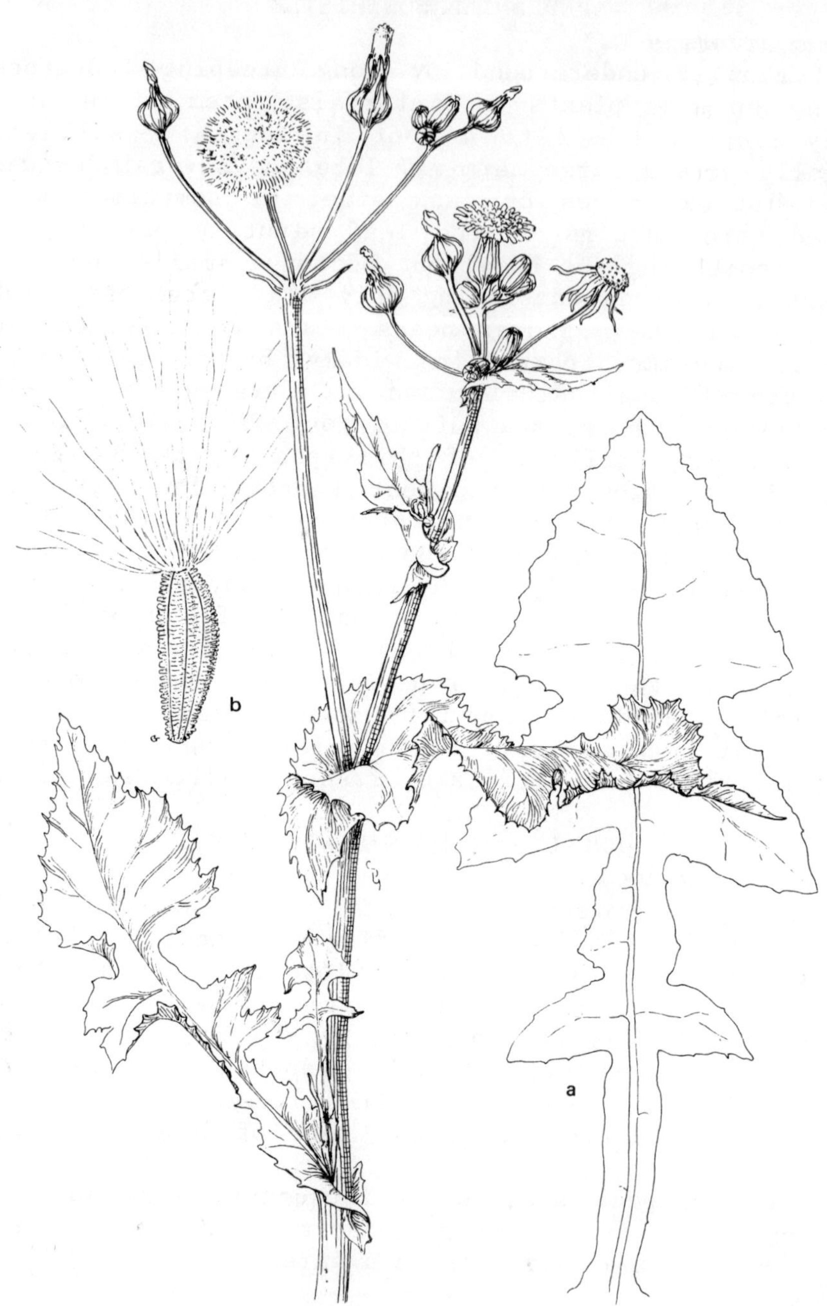

ANNUAL SOWTHISTLE
Sonchus oleraceus x 2/3
a. basal leaf x 2/3
b. fruit x 10

BRISTLY HAWKSBEARD

Crepis setosa Hall. f.

Annual; stems one or several from the base, branching, 1 to 2 1/2 feet tall, somewhat hairy; stem leaves broadest at the base, coarsely toothed or lobed; heads yellow, generally numerous, small, spreading, bracts of the head coarsely and stiffly bristly; fruits long-beaked.

Bristly Hawksbeard was introduced from Europe, and has become a common and abundant weed of lawns, gardens and waste places west of the Cascades. It now replaces to some degree Smooth Hawksbeard, and appears to be more aggressive than that species.

BRISTLY HAWKSBEARD
Crepis setosa x 1/2
a. fruit x 2 1/2

SMOOTH HAWKSBEARD
Crepis capillaris x 2/3
a. fruit x 15

SMOOTH HAWKSBEARD

Crepis capillaris (L.) Wallr.

Annual or biennial; stem 1/2 to 2 feet tall, one to several from the base, branching, ridged, sometimes purplish or reddish below; leaves lobed or narrowly toothed, nearly smooth; heads 1/2 inch wide or less, numerous, borne at the ends of slender branches; flower heads yellow, enclosed by a circle of narrow bracts, these inconspicuously hairy; fruit 1/12 inch long, pale brown, narrow, 10-ribbed, with a terminal tuft of soft white hairs.

This species has long been common west of the Cascades and so abundant in wasteland as to color large areas yellow. It gives trouble in gardens, pastures, and fields of western valleys, and occasionally is reported as a nuisance in localized areas east of the Cascades as well. It is a native of Europe.

NIPPLEWORT

Lapsana communis L.

Annual; stem erect, simple or branched above, 4 inches to 4 feet in height; leaves broadly triangular to roundish, the lower often with one to several small remote lobes, long stalked, leaves narrowing upward and borne on progressively shorter stalks, leaf blades irregularly toothed, soft hairy to smooth; heads several to many, bracts firm, erect, flowers yellow; fruit 1/8 to 1/4 inch long, striate.

Naturalized from Eurasia and now a common weed, particularly west of the Cascades, usually in fields and disturbed habitats.

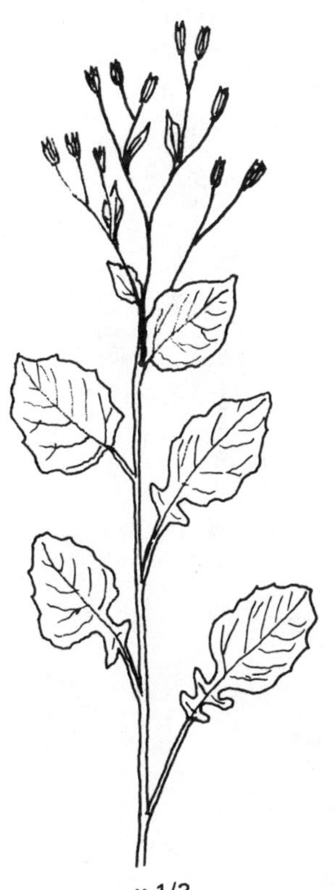

x 1/3

BLUE LETTUCE
Lactuca pulchella x 2/3
a. leaf x 2/3
b. fruit x 3

BLUE LETTUCE

Lactuca pulchella (Pursh) DC.

Perennial, with a long rootstock; plant bluish green, smooth; stem 1 1/2 to 3 feet tall, erect or nearly so, one to several from the base, leafy; upper leaves long, narrow, their midribs pale; heads in loosely-branched clusters, blue or purple, fruit elongated, short-beaked, bearing a terminal tuft of soft white silky unbranched hairs.

Blue Lettuce is frequently reported as a weed in cultivated fields, particularly east of the Cascades. Its deep-seated rootstock enables it to spread rapidly, and makes it somewhat difficult to control. It is a native of North America.

PRICKLY LETTUCE

Lactuca serriola L.

Annual, winter annual, or biennial; stem 1 1/2 to 6 feet tall, erect, branched above, more or less prickly below, the entire plant bluish green; lower leaves sometimes lobed, upper leaves sharply toothed, lobed or not, their bases clasping the stem by ear-like projections, midribs of the leaves generally with a line of sharp prickles; flower heads pale yellow, on slender stalks; buds of heads long; fruit slender, parachute-like, 10-ribbed, long-beaked, tipped by a circle of white hairs.

Prickly Lettuce has become very abundant in the Pacific Northwest. It is native to Europe but is now known on almost every continent. Distribution of the weed is accomplished by air currents which carry the parachute-like fruits high and wide.

WILLOWLEAF LETTUCE

Lactuca saligna L.

Biennial; stem erect; leaves not lobed or with a few spreading lobes at the base, narrowly arrow-shaped, the ear-like basal projections often toothed; flower heads yellow; fruit slender, long-beaked.

Another native of Europe, this species has become rather sparingly distributed in northwestern and southern Oregon. It is, in some cases, locally abundant, and a pest in cultivated fields. Like other species of lettuce, it is distributed by its parachute-like fruits and can readily cover large areas in a short time.

PRICKLY LETTUCE
Lactuca serriola x 2/3
a. fruit x 3

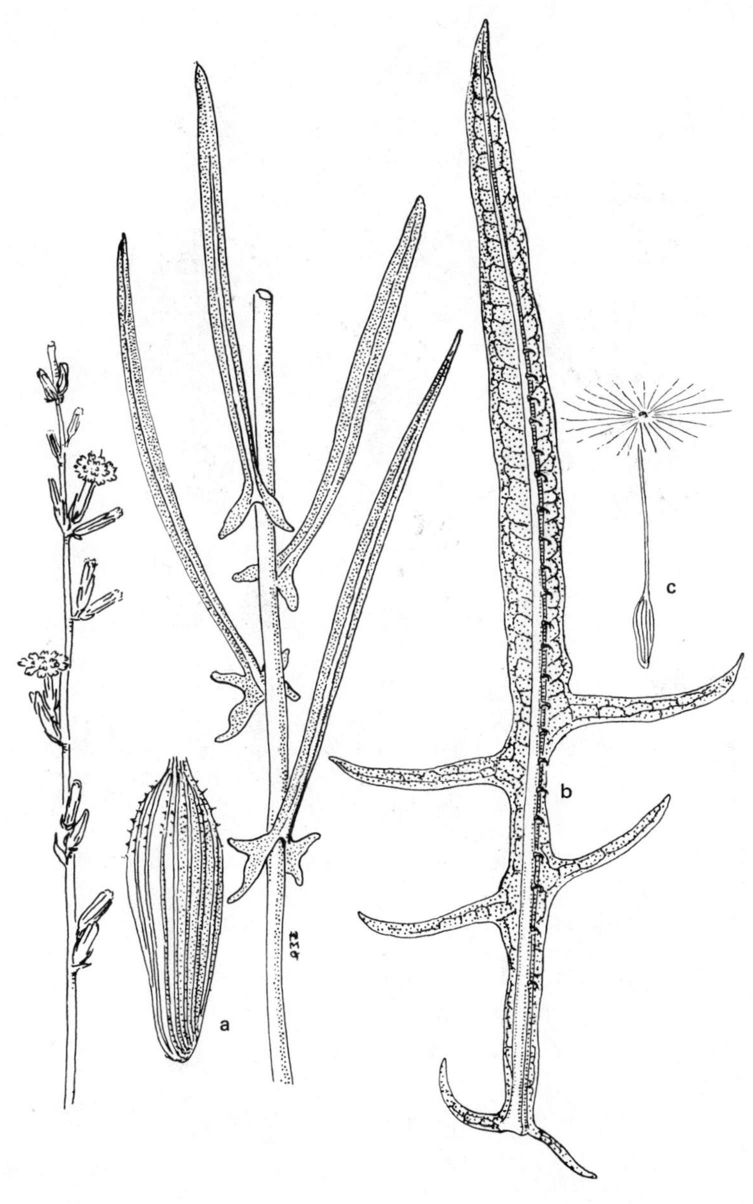

WILLOWLEAF LETTUCE

Lactuca saligna x 2/3

a. fruit with beak and pappus removed x 12

b. lower leaf x 2/3

c. fruit x 3

CHICORY
Cichorium intybus x 2/3
a. root x 2/3
b. fruit x 6
c. basal leaf x 2/3

CHICORY

Cichorium intybus L.

Perennial; stem 1 to 8 feet tall, stiff, ridged, grayish-green, branches few but spreading; leaves rough, basal leaves 2 to 10 inches long, stalked, toothed or lobed, borne in a rosette; stem leaves becoming gradually smaller above, stalkless, with 2 ear-like lobes clasping the stem; heads deep blue (rarely pink or white), 1 1/2 inches or less in diameter, spreading; fruit 1/12 to 1/8 inch long, yellow, brown, or black, obscurely angled or ribbed, tipped by a crown of minute scales.

This attractive wayside plant, exhibiting one of the clearest blue shades known in flowers, is a native of the Mediterranean region so far as known, though it now is found on every continent. Esteemed in Europe as a salad plant and for cooking, as well as for a coffee substitute, it has accompanied the white man on his travels throughout the world. Its introduction into new lands generally was deliberate rather than accidental.

In the late nineteenth and early twentieth century, attempts were made to establish it in America as a farm crop, the roots to be used in the manufacture of a coffee substitute or adultrant. Since then, this species has become a common inhabitant of roadsides and waste places and occasionally is troublesome in cultivated fields.

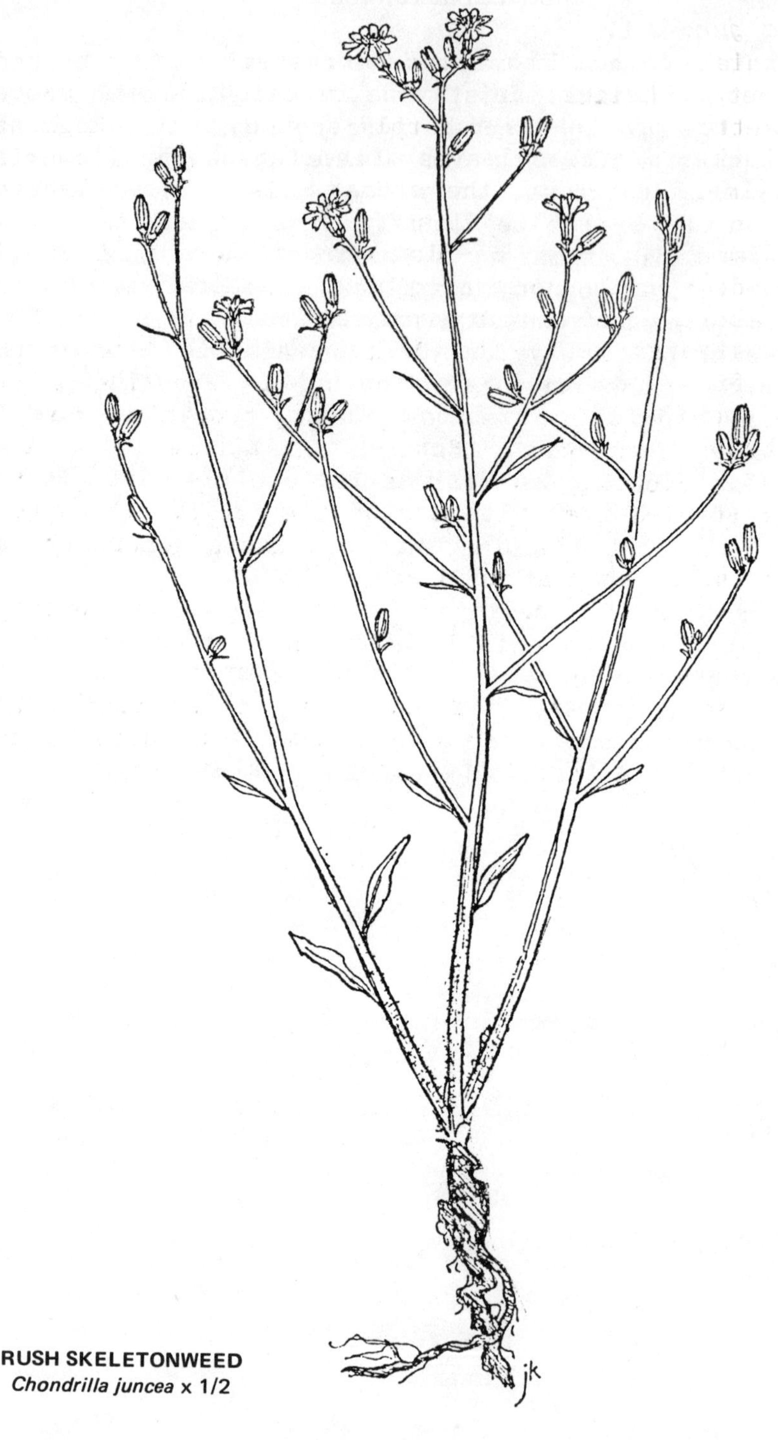

RUSH SKELETONWEED
Chondrilla juncea x 1/2

RUSH SKELETONWEED

Chondrilla juncea L.

Perennial from a stout taproot; stems much branched, up to 4 feet in height, bristly-hairy below, smooth above; basal rosette of leaves sharply-toothed, the segments directed backward, these leaves withering as the flowering stems develop; leaves of the stems entire, linear; heads scattered on the nearly leafless branches; heads 3/4 of an inch in diameter, 7- to 15-flowered, flowers yellow, all strap-shaped; fruit several-ribbed, smooth below but roughened above by tiny scaly projections from among which comes the slender beak, the beak terminated by numerous soft white fine bristles.

This species is an introduction from Europe. It is occasional in the Northeastern United States and on the West Coast it occurs in California, Idaho, eastern Washington and in Douglas County, Oregon. Because, once established, it can spread rapidly and is difficult to control, it can be expected to extend its range in Oregon, from the Douglas County infestation and to appear on the northeastern and eastern borders of the state from Washington and Idaho populations.

Rush Skeletonweed appears to prefer well-drained soil and does well along roadsides in cultivated areas and on rangeland and is a particular threat to dryland agriculture.

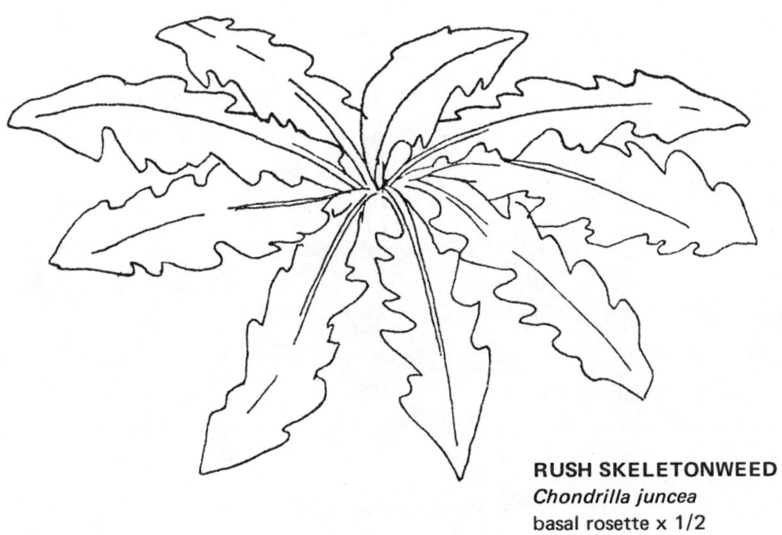

RUSH SKELETONWEED
Chondrilla juncea
basal rosette x 1/2

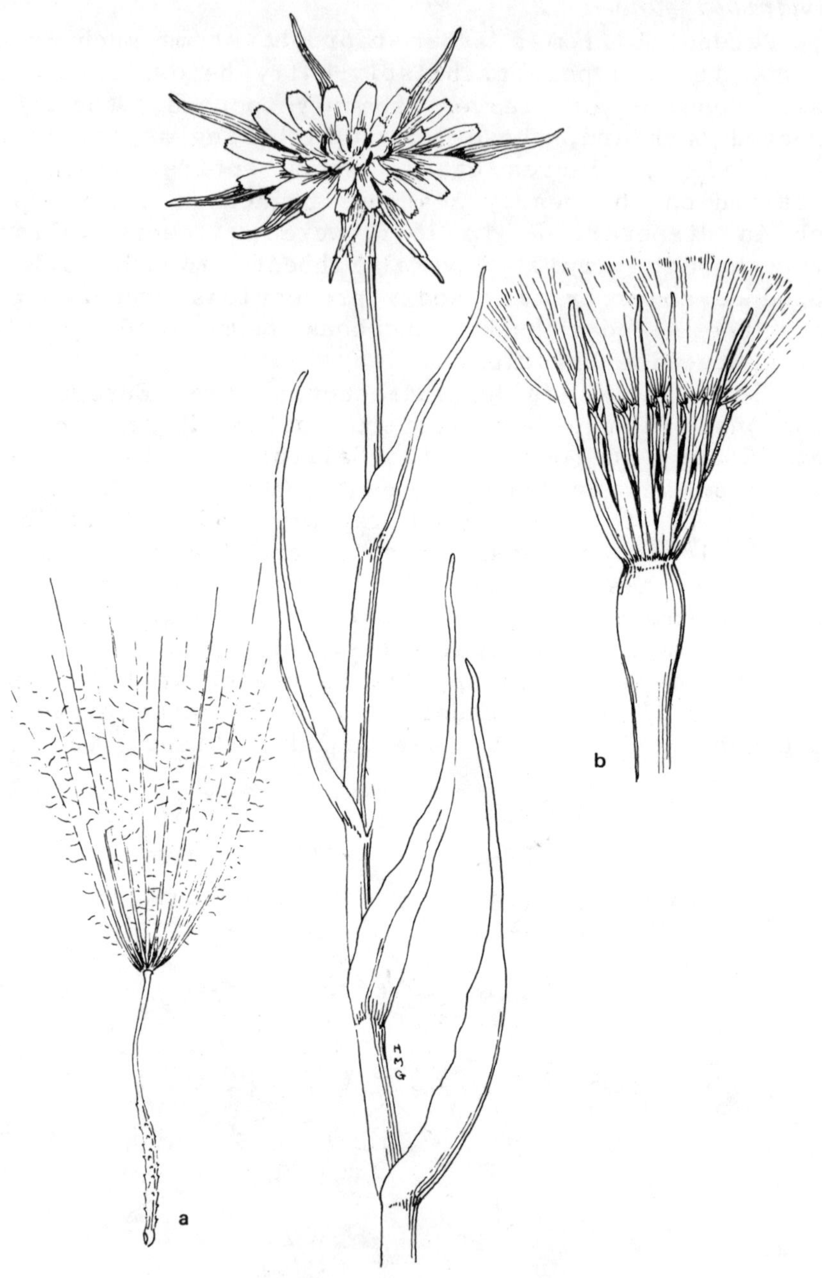

WESTERN SALSIFY
Tragopogon dubius x 1
a. fruit x 1 1/2
b. head (in fruit) x 1

WESTERN or YELLOW SALSIFY

Tragopogon dubius Scop.

Biennial; stem 1 to 3 feet tall, stiff, thickened below the flower head; leaves clasping; flowers yellow; fruit slender, long-beaked, the beak tipped by a circle of stiff plume-like bristles.

A native of Europe, this species has become well-established in areas east of the Cascades, and is occasionally found west of this mountain range. Another yellow-flowered species, *Tragopogon pratensis* L., Meadow Salsify, occurs here, also. The stalk of this species is not swollen below the flower head.

Purple-flowered or Common Salsify, *Tragopogon porrifolius* L., a garden vegetable, occasionally is encountered along roadsides and on wasteland where it has escaped from cultivation.

SPINY COCKLEBUR

Xanthium spinosum L.

Annual; stems spreading or erect, 2/3 to 2 feet long; leaves densely covered below with short white hairs, white-veined above, blade 1 to 3 inches long, narrow, with generally 2 short basal lobes or teeth, and bearing a stiff 3-forked spine at junction with the stem; "bur" bearing a beak and prickles. Spiny Cocklebur came to us from Europe. It is sparingly established in dry areas of the Pacific Northwest.

x 2/3

HEARTLEAF or BROAD-LEAVED COCKLEBUR

Xanthium strumarium L.

Annual; stem more or less branched, 2/3 to 1 1/2 feet long, erect or spreading, often with purple dots and lines; leaves broadly heart-shaped, more or less lobed, roughened on the surface; "burs" with conspicuous curved beaks and stiff hooked bristles.

Two varieties of this species occur in our limits. Heartleaf Cocklebur, though now world-wide in distribution, probably originated in North America. This species has proved troublesome, not only as a weed of farmyards and cultivated fields, but also as a source of poisoning of grazing animals and domestic fowls.

ANNUAL BURSAGE or BURWEED

Ambrosia acanthicarpa Hook.

Annual; stem 1 to 3 feet tall, branching; leaves more or less lobed or divided, the lower long-stalked; staminate heads borne in terminal racemes, the pistillate heads spiny and borne singly or in 2's or 3's in the leaf axils. Some plants are often chiefly staminate or chiefly pistillate.

This species and Ragweed considerably resemble each other in vegetative characters, but Bursage is more conspicuously stiff-hairy, particularly on the young leaves, and the leaves are never paired. Fruits of the two species are readily distinguished, those of Ragweed, bearing several wart-like projections in usually a single circle; while those of Bursage are covered by several circles of sharp spines.

Annual Bursage is found principally east of the Cascades in dry areas.

x 1

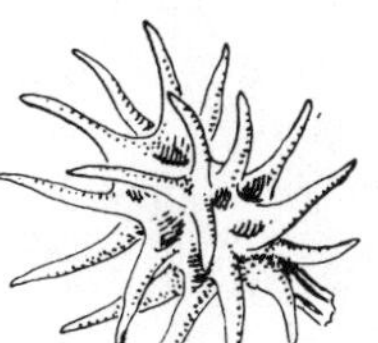

x 4

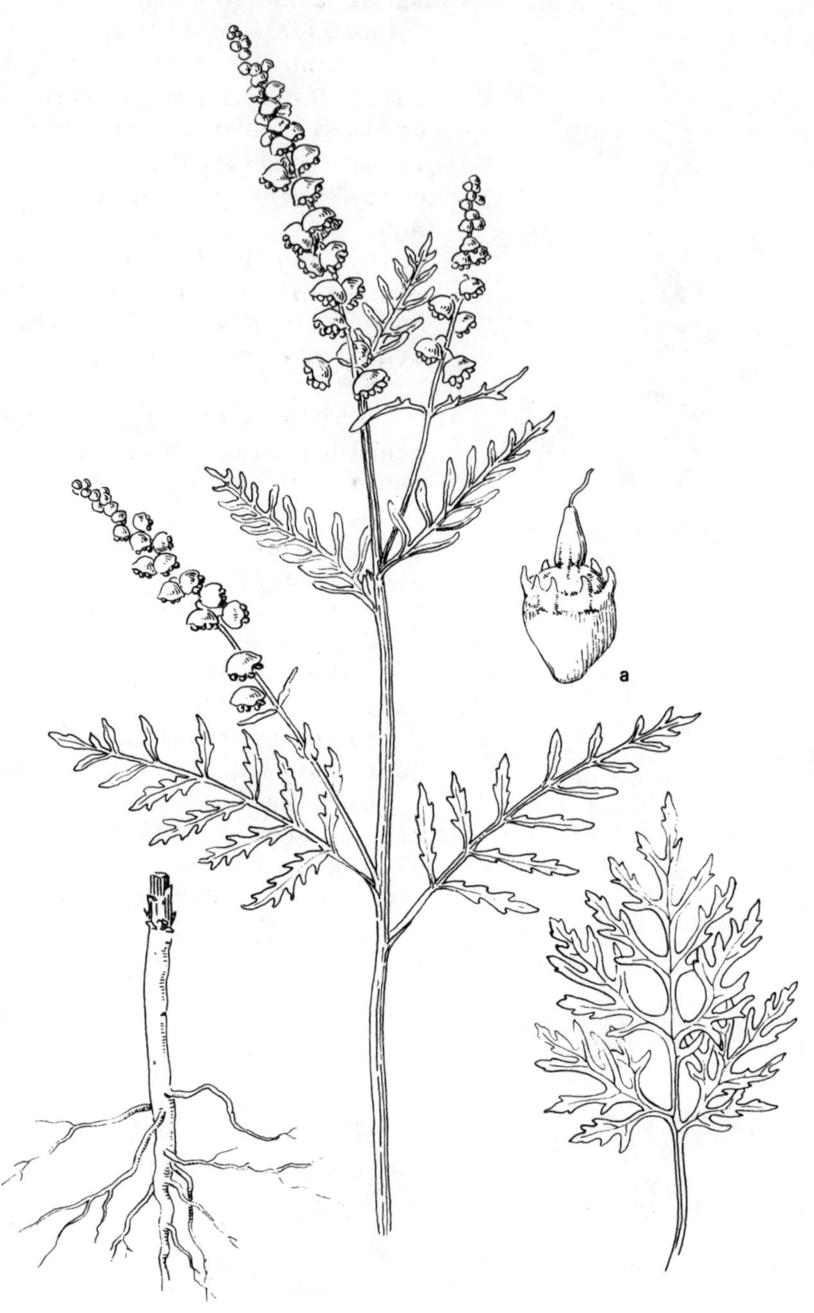

COMMON RAGWEED
Ambrosia artemisiifolia x 1
a. fruit x 5

COMMON RAGWEED

Ambrosia artemisiifolia L.

Annual; stem erect, 1/2 to 6 feet tall, simple or branched, smooth to rough-hairy; leaves generally paired near base of the stem, alternate above, or sometimes either paired or alternate throughout, stalked, the blades deeply lobed or more frequently divided into leaflets, these again once or twice divided or lobed, hairy; flower heads small, of 2 kinds, the staminate heads nodding, borne in long racemes, the pistillate heads borne in the upper leaf axils; fruit enclosed in a hard covering bearing several wart-like structures, principally in a circle.

Within comparatively recent years this species of Ragweed has spread rather alarmingly in several unconnected areas east and west of the Cascades. The plant is native to the eastern United States where it exists in several somewhat distinct forms. That which occurs in our area appears for the most part to be variety *elatior* (L.) Desc.

Inasmuch as this species is not only a pernicious pest in cultivated fields, but is said to be responsible for 85% of autumn cases of hayfever in the eastern and southern United States, it is needless to emphasize its undesirability as an inhabitant of our area. By adding it to our general plant population, we may lengthen our hayfever season by several months, since our most disturbing pollens west of the Cascades at present are largely those of spring and very early summer.

Giant Ragweed (*Ambrosia trifida* L.) and annual which may reach a height of 15 feet or more; and Western or Naked-spiked Ragweed (*Ambrosia psilostachya* DC.), a lower plant which is generally perennial by creeping rootstocks, occur more or less frequently east of the Cascades in the states here included. All are heavy pollen producers and contribute to disturbances involving allergy.

POVERTYWEED
Iva axillaris x 1/2

POVERTYWEED or DEATHWEED

Iva axillaris Pursh

Perennial by a spreading branching system of rootstocks and roots; stem erect, several inches to 2 feet tall, leafy; leaves small, narrowly or broadly oblong, obscurely 3-veined, thickish, lower leaves generally paired on the stem, the others alternate, the leaves stalked to nearly stalkless; flower heads borne singly in the leaf axils, downward-turned; fruit 1/12 to 1/10 inch long, broadest at the apex, more or less covered by minute golden glands.

In being a native, Povertyweed differs from most other weeds in our area. Nevertheless it may become very troublesome in an area subjected to abnormal conditions such as overgrazing or cultivation, in either of which the natural balance is disturbed. It thrives in alkaline soil but can adapt itself to other soil types. The plants are more or less gland-dotted and have a strong disagreeable odor which, together with an abundance of pollen, makes them an irritant to hayfever sufferers. The ability to spread underground and form large surface colonies, thus crowding out better plants, renders this species a noxious weed in pastures and cultivated fields, and difficult to control. It is reported troublesome in a number of counties east of the Cascades.

MARSH ELDER

Iva xanthifolia Nutt.

Annual; stem 2 to 6 feet tall, generally unbranched below; leaves (at least the lower) paired on the stem, the uppermost sometimes alternate, long-stalked, the blades 2 to 6 inches long, short-hairy, the bases broad and rounded or heart-shaped, the apices long-pointed, the margins toothed or lobed; flower heads small, numerous, borne in large terminal branching clusters; fruit broad or narrow, somewhat angled, broadest at the apex, about 3/16 inch long.

Though spectacular because of its large size, this species is less troublesome than Povertyweed, but possesses the same tendency to broadcast its irritating pollen. The plant is known in central and southeastern Oregon, Idaho, and Washington.

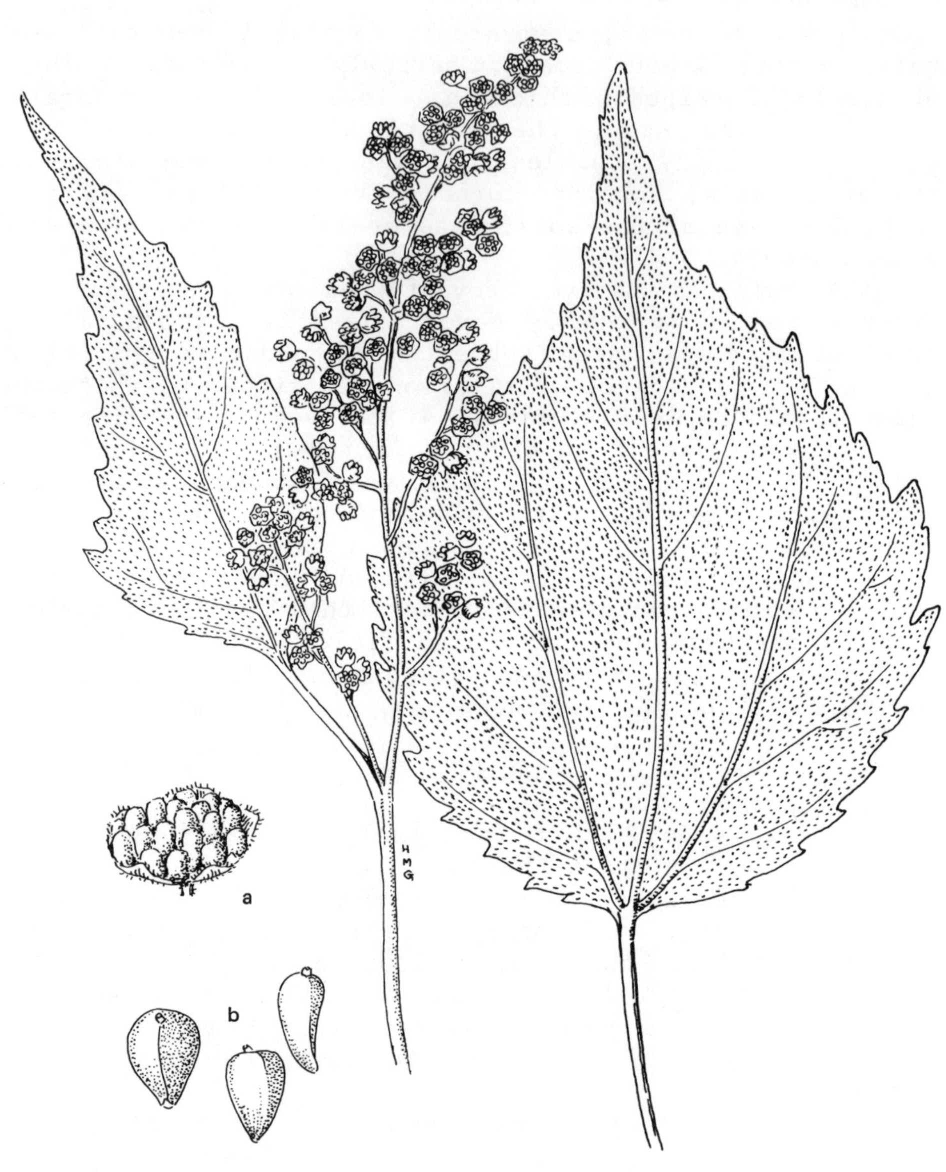

MARSH ELDER
Iva xanthifolia x 4/5
a. head of flowers in bud x 4
b. fruits x 4

SUNFLOWER

Helianthus annuus L.

Annual up to 8 feet in height or sometimes taller; leaves rough, more or less triangular in shape, or the lower somewhat heart-shaped, stalked, usually with 3 main veins from near the base of the blade, the margins coarsely and irregularly toothed; heads terminal or in the axils of the upper leaves; ray flowers bright yellow, disc flowers brownish; fruit nearly wedge-shaped, tipped by 2 short, early deciduous, scales.

This species has long been cultivated for its edible seeds and several horticultural forms have been developed with large heads. In the wild, however, this species is weedy; in our area it is found mainly east of the Cascades.

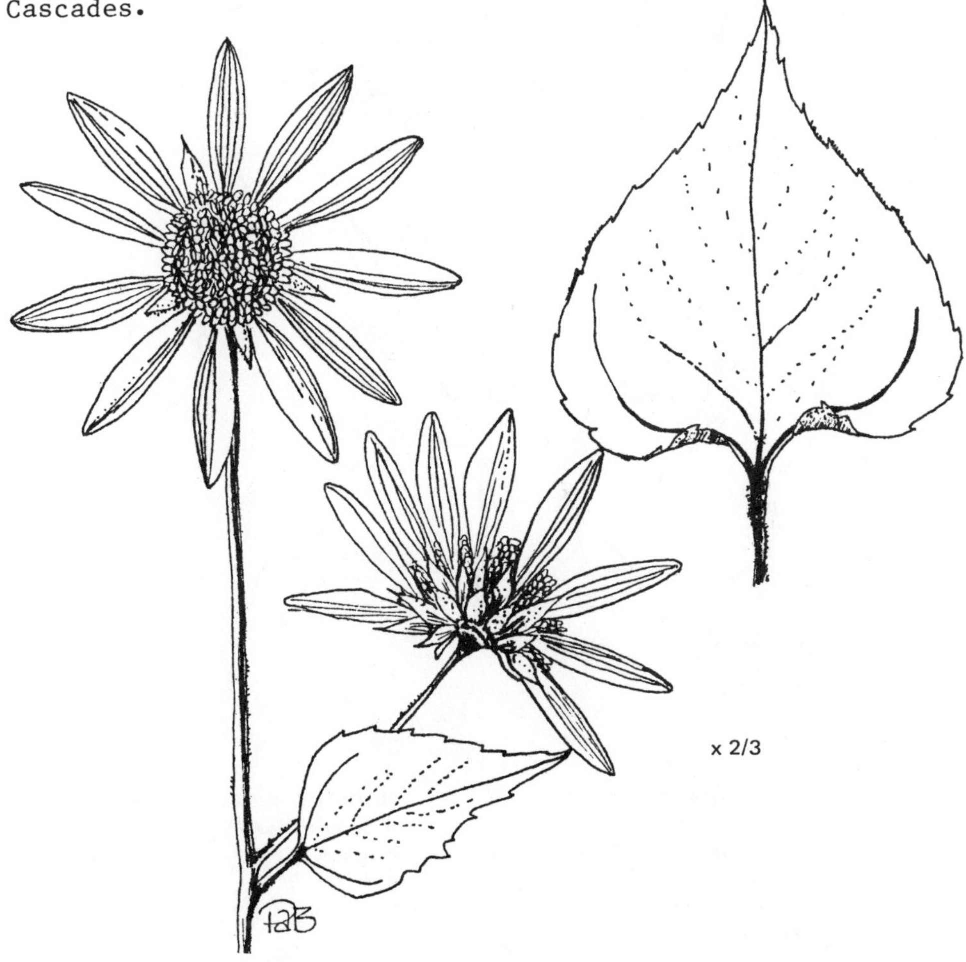

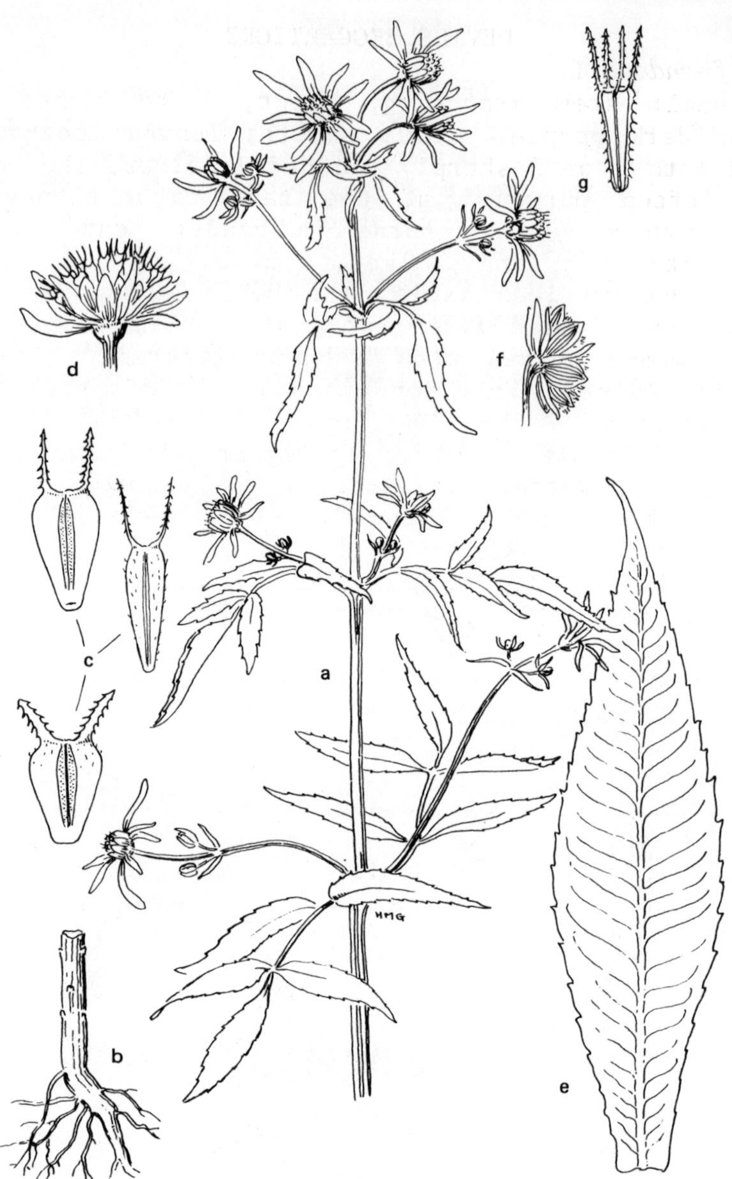

DEVILS BEGGARTICKS
Bidens frondosa
a. upper portion of plant x 2/3
b. root x 2/3
c. fruits x 2
d. head (in fruit) x 1

NODDING BEGGARTICKS
Bidens cernua
e. leaf x 2/3
f. head x 2/3
g. fruit x 2

DEVILS BEGGARTICKS

Bidens frondosa L.

Annual; stem branching, erect, 1 to 3 feet tall, somewhat dark purple, nearly smooth; leaves long-stalked, divided into 3 or 5 sharply toothed leaflets, the terminal leaflet often narrowed at the base to a slender long stalk; flowers yellow, borne in heads; heads generally many, short- or long-stalked, borne at the ends of branches and in leaf axils, the heads surrounded at the base by several leaf-like bracts of varying length; fruit narrowly wedge shaped with 2 horns generally with stiff backwardly turned barbs, or occasionally with erect stiff bristles.

Several other species of Beggarticks occur in our area. Tall or Western Beggarticks, *Bidens vulgata* Greene, resembles the preceding species but is generally larger, with longer leaf-like bracts below the flower head, and larger fruits. Occasional plants reaching a height and horizontal spread of 7 feet or more have been seen. Nodding Beggarticks (*Bidens cernua* L.) has undivided, toothed leaves and nodding heads, with generally 4-horned fruits.

The beggarticks naturally inhabit moist ground, and therefore may give trouble in irrigated fields. They are rather common on bottom lands, and may be transferred to gardens through river loam. The barbed and horned fruits have the disagreeable habit of becoming attached to clothing and to the wool and fur of animals.

CLUSTER TARWEED
Madia glomerata
a. upper portion of plant x 2/3
b. fruit of disk flower x 15
c. fruit of ray flower x 15

SHOWY TARWEED
Madia elegans
d. upper portion of plant x 2/3
e. fruit x 15

CLUSTER or STINKING TARWEED

Madia glomerata Hook.

Annual; stem erect, 1 to 2 feet tall, leafy, simple or branching, entire plant very sticky-glandular, disagreeably scented, hairy; leaves narrow; flowering heads inconspicuous, clustered at the apex of the stem, at ends of branches and in leaf axils, always partially enclosed by leaves; rays few, short, yellow; fruit narrow.

This is one of our most common tarweeds and often forms dense stands in dry fields of both valley and hillside. Its obnoxious odor and its sticky surfaces which catch and hold dust, only to deposit it later on the clothing of persons walking through it, make it one of our most annoying weeds so far as direct contact is concerned. The "tar," when combined with dust, is difficult to remove from clothing, the hair of dogs, and the wool of sheep.

The plant is said to be poisonous to stock, but under normal conditions it is avoided by grazing animals. This fact, however, contributes to its persistence and increase in closely grazed pastures where forage plants of value are eaten off.

SHOWY TARWEED

Madia elegans D. Don

Annual; stem erect, 1 1/2 to 2 1/2 feet tall; upper part of the plant somewhat sticky-glandular, stiff-hairy; leaves narrow, stiff-hairy; flower heads borne at the ends of slender branches; rays large, showy, toothed, pure yellow, or reddish-brown at the base; fruit somewhat flattened, slightly curved, broad.

This is a native tarweed, conspicuous in fields along roadsides, but scarcely to be considered a pest in comparison with other species.

CHILEAN TARWEED

Madia sativa Mol.

Annual; stem stout, 1 to 5 feet tall; heads generally broad, often in dense terminal clusters not surrounded by leaves; rays short, inconspicuous, fruits of ray flowers broad, flattened; of the disk flowers, narrow, 4-angled.

This species, in common with Cluster Tarweed, is abundant in many parts of western Oregon. It is said to be a native of South America, having spread northward along the coast. It, too, has a disagreeable odor and an abundance of sticky glands.

Several other species of tarweed are found in the Northwest, a few considerably smaller than those described. For the most part they are summer-flowering, ill-scented, sticky plants of dry areas. Their fruits are a favorite food of certain birds, including goldfinches, which visit them in flocks during late summer or fall.

CHILEAN TARWEED
Madia sativa x 3/4

a. involucral bract x 4
b. fruit from ray flower x 8
c. fruit from disk flower x 8

SPIKEWEED

Hemizonia pungens (H. & A.) T. & G.
(=*Centromadia pungens* Greene)

Annual; stems 1 1/2 to 3 or more feet tall, pale straw-colored or whitish, stiff, glandular; basal leaves yellowish green, several inches long, with narrow lobes, stem leaves 1/2 inch or less long, needle-pointed, many bearing dwarfed vegetative branches or short flowering branches in their axils; heads borne at tips of short leafy branches, yellow, about 1/3 inch broad; fruit 1/16 inch long, with wart-like projections and a short beak.

This is a native Californian species which is being reported rather frequently from eastern Oregon. It thrives on alkaline soil and, by seeding freely, densely covers areas favorable to it. Because of its rough spiny character it is a disagreeable weed, particularly in pastures.

SPIKEWEED
Hemizonia pungens x 1/2
a. basal leaf x 1/2
b. fruit x 15

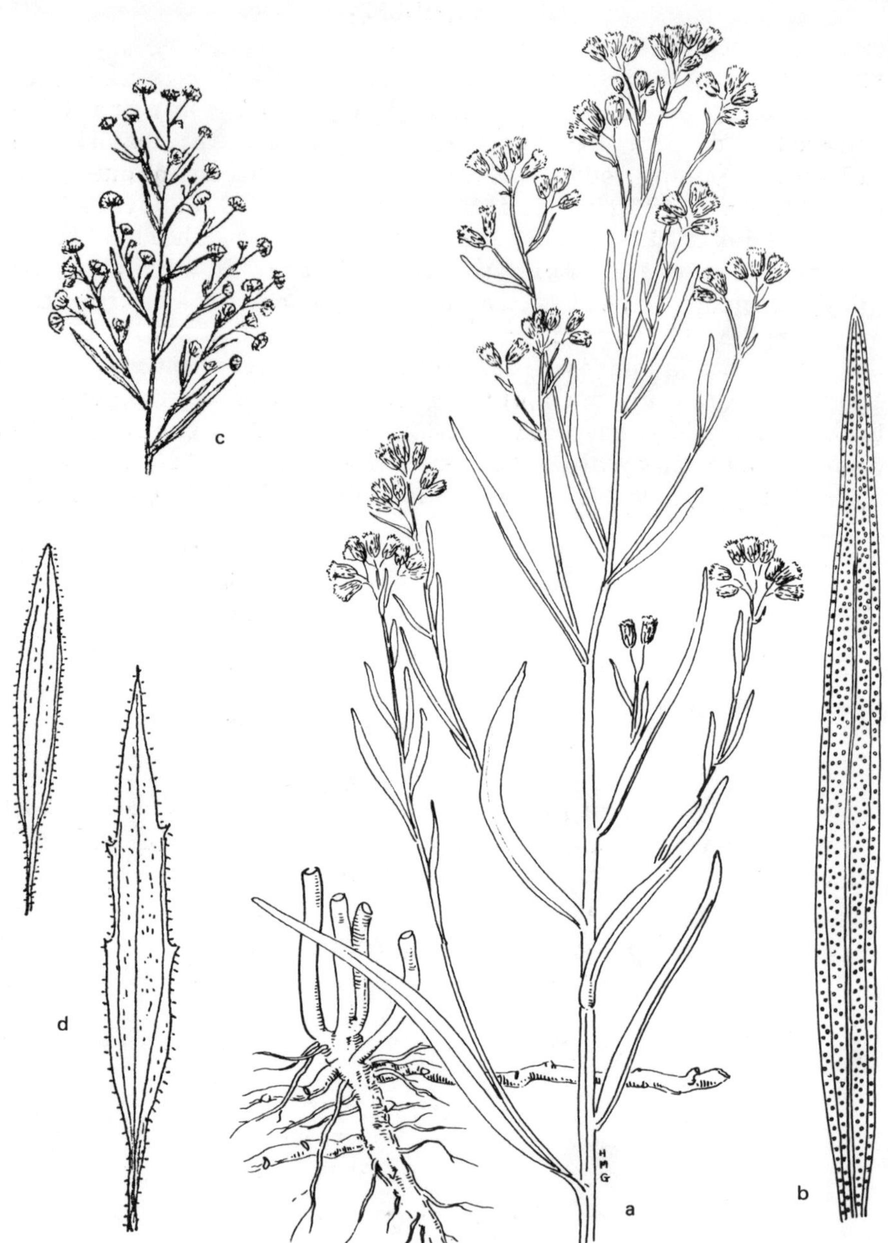

HORSEWEED
Conyza canadensis
c. portion of inflorescence x 2/3
d. leaves x 1

WESTERN GOLDENROD
Solidago occidentalis
a. plant x 2/3
b. leaf x 1 1/3

WESTERN GOLDENROD

Solidago occidentalis (Nutt.) T. & G.

Perennial, spreading by thick creeping rootstocks; stem 2 to 3 feet tall, woody, dark green, one or several from the base, branching above, branches erect; leaves numerous, very narrow, 3-veined, minutely punctate; heads generally numerous, very small, gummy, with minute rays; fruit hairy.

This native species of Goldenrod has proved somewhat troublesome in irrigated fields and along ditch banks. Its rapid method of underground propagation gives it an advantage over most field crops and renders control difficult.

Another species found almost exclusively east of the Cascades is Giant Goldenrod (*Solidago gigantea* Ait. var. *serotina* (Kuntze) Cronq.). It differs from Western Goldenrod in its wide-spreading flowering branches which bear golden yellow heads in showy sprays.

The dominant species of valleys west of the Cascades is Canada Goldenrod (*Solidago canadensis* L.). It somewhat resembles Giant Goldenrod in leaf characters, but its flowering branches are generally erect and compactly arranged, forming a dense yellow plume.

The two latter species, though common and frequently abundant, are rarely reported as troublesome.

HORSEWEED

Conyza canadensis (L.) Cronq.
(=*Erigeron canadensis* L.)

Annual; stems 2/3 to 5 feet in height, bristly-hairy to nearly smooth; lower leaves up to 4 inches long and 3/8 inch wide, stalked, soon withering; leaves of the stem narrower and not stalked, usually with a fine fringe of hairs on the margins; heads small and numerous, the rays white and inconspicuous, disc flowers yellow; fruit minute, tipped by slender white bristles.

The native variety is var. *glabrata* (Gray) Cronq. in which the stems are nearly smooth, or with just a few spreading hairs. Variety *canadensis* is bristly-hairy and is presumably introduced. Both varieties are common weeds in pastures, roadsides, cultivated areas and disturbed wasteland.

RUBBER RABBIT-BRUSH
Chrysothamnus nauseosus x 2/3
a. floret x 4

RUBBER or GRAY RABBIT-BRUSH

Chrysothamnus nauseosus (Pall.) Britt.

Perennial, strongly-scented, shrubby below, much branched; stems erect or sometimes spreading, 3/4 to 6 feet long, more or less white-woolly or white-silky, this covering sometimes more or less disappearing except in the leaf axils; leaves numerous, slender, 3/4 to 2 inches long, white-woolly or white-silky; flower heads yellow, borne in large flat-topped or round-topped clusters; fruit hairy, topped by a tuft of microscopically barbed bristles.

This native species of the arid sections east of the Cascades has become a pest on certain of the stock ranges. It is commonly associated with Sagebrush but, because of its ability to crown-sprout after spraying or burning, it is more aggressive and less effectively controlled. In areas where Sagebrush originally predominated over Rabbit-brush but has been eliminated by surface burning or by the use of herbicides, Rabbit-brush tends to recover and spread, sometimes becoming an even greater menace than was the Sagebrush. This behavior of the weed is causing serious concern in areas which, after clearing, have been seeded to Crested Wheatgrass but subsequently have become invaded by Rabbit-brush.

Another species with the same disagreeable habit of regenerating from the crown is Douglas or Sticky Rabbit-brush, *Chrysothamnus viscidiflorus* (Hook.) Nutt., which occurs in similar situations but differs in several respects. In its more common form the leaves are generally wider than those of Rubber Rabbit-brush, and more or less distinctly 3-veined longitudinally; the stems and leaves are either smooth or short-hairy but never woolly or silky; and a sticky secretion is usually found at the bases of leaves and branches, often trickling down the stem.

COMMON BURDOCK

Arctium minus Benth.

Biennial; plant bushy-branched, 1 1/2 to 10 feet tall, more or less woolly at first; leaves large (sometimes reaching 1 1/2 feet in length), broad, roughened, toothed or wavy-margined, woolly beneath at least when young; heads generally numerous, very short stalked, clustered along flowering branches, generally less than 1 inch broad, including spread of the spines; spines numerous, slender, stiff, often purplish, hooked at the ends; flowers purple.

This is a European species rather commonly distributed, and constituting a pasture weed. The bur-like heads, covered by hooked bristles, become disagreeably attached to clothing and to the coats of animals.

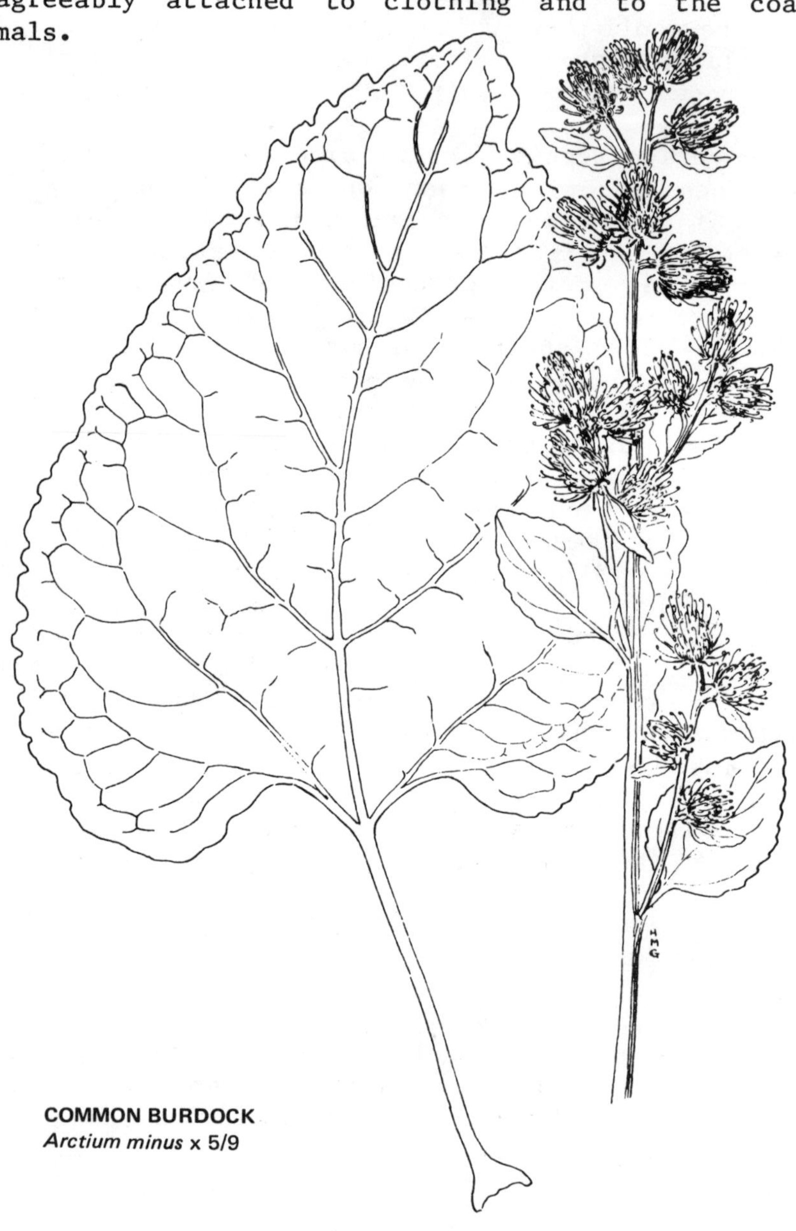

COMMON BURDOCK
Arctium minus x 5/9

ENGLISH or EUROPEAN DAISY

Bellis perennis L.

Perennial; leaves basal, prostrate or spreading, nearly smooth or loosely hairy, entire-margined or toothed, broad above, narrowed at the base to a long stalk; flower heads white or pinkish with yellow centers, erect, long-stalked, the stalks generally exceeding the leaves in length.

The English Daisy is, as its name implies, a native of Europe. Introduced as a garden plant into many other parts of the world, it has in many cases escaped bounds and become naturalized. It is well known west of the Cascades as a lawn weed--though sometimes purposely introduced into lawns for contrast and color--and in pastures and waysides along the coast. Its tendency to form a solid stand makes it unwelcome in pastures where it effectively crowds out the grasses.

x 4/5

BLACK KNAPWEED

Centaurea nigra

a. upper portion of plant x 2/3
b. floret x 2
c. involucral bract x 4
d. fruit x 6

MEADOW KNAPWEED

Centaurea pratensis

e. upper portion of plant x 2/3
f. two types of florets x 2
g. involucral bract x 4
h. fruit x 6

BROWN KNAPWEED

Centaurea jacea

i. involucral bract x 4
j. fruit x 6

BLACK KNAPWEED

Centaurea nigra L.

Perennial; stems leafy, branching, rough-hairy, 2 to 4 or more feet long, weakly erect, often with the bases prostrate and rooting at several nodes; leaves short-hairy, somewhat grayish green, the margins entire or toothed, the lower leaves long-stalked, blades oblong, the upper leaves stalkless; heads often many, about 1 inch high, the bulbous base at first nearly globose, covered by purplish black fringed bracts, the flowers purplish-red; fruit with a circle of short scales at the apex.

A native of Europe, Black Knapweed has reached North America and New Zealand. The fairly rapid spread of this species has caused some concern to farmers.

BROWN KNAPWEED

Centaurea jacea L.

Brown Knapweed, likewise a perennial, somewhat resembles Black Knapweed. But the lower leaves are generally lobed; the unopened heads slightly elongated; the bracts brown, not fringed with hairs but with thin papery margins; the outer flowers of the head enlarged and spreading raylike; and the fruit without a circle of scales at the apex.

This species, like Black Knapweed, is European and is well established in western America. It, too, has purplish-red flowers. In certain localities west of the Cascades where it has become somewhat abundant, it is utilized for pasture; but it is relatively coarse and harsh and, where it has escaped bounds, has become troublesome.

MEADOW KNAPWEED

Centaurea pratensis Thuill.

In the Willamette Valley, and occasionally in other parts of our area, occurs another purple-flowered knapweed which is known as Meadow Knapweed. It is considered to be a hybrid between the two preceding species, both of which it resembles in certain characters. In foliage, color of bracts, in enlarged outer florets, and in the usual absence of appendages from the fruit, it is like Brown Knapweed, but its bracts are deeply fringed as in Black Knapweed.

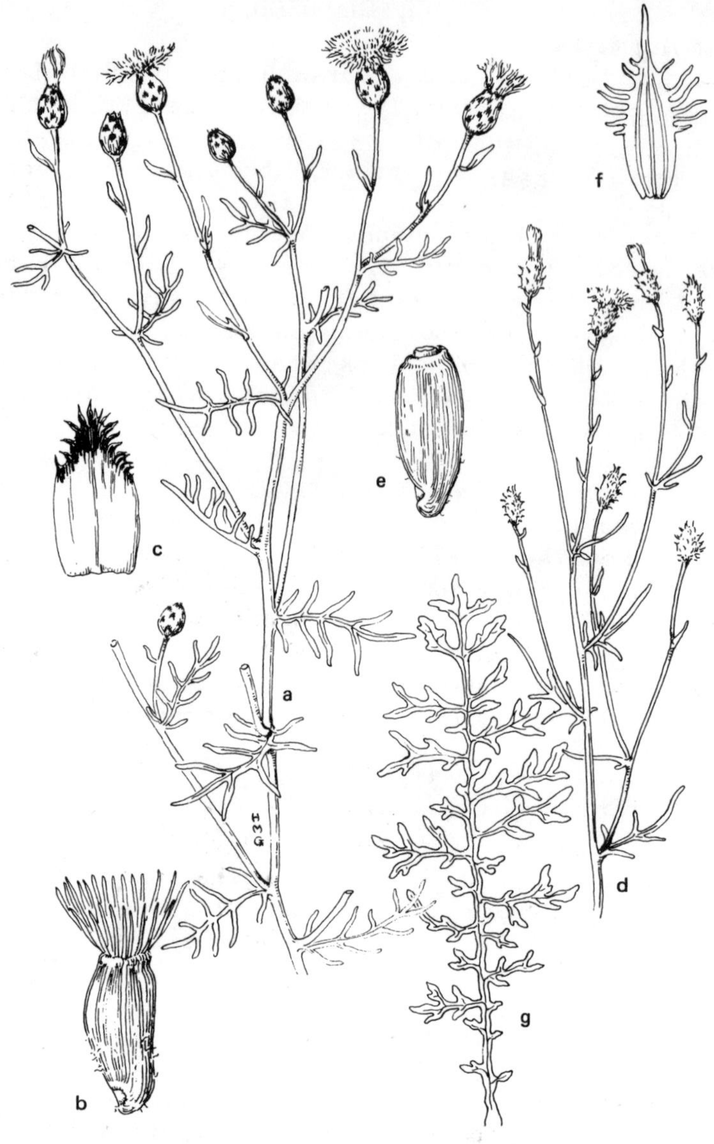

SPOTTED KNAPWEED
Centaurea maculosa
a. upper portion of plant x 2/3
b. fruit x 8
c. involucral bract x 6

DIFFUSE KNAPWEED
Centaurea diffusa
d. upper portion of plant x 2/3
e. fruit x 8
f. involucral bract x 6
g. lower leaf x 2/3

SPOTTED KNAPWEED

Centaurea maculosa Lam.

Biennial or short-lived perennial; generally somewhat loosely woolly, 1 to 3 feet tall, stiffly branching, the branches more or less erect, rough, ridged; leaves except the uppermost, divided featherwise into narrow segments; heads purple-flowered, with a spread of 3/4 inch or less, the bulbous basal structure elongated, its bracts black tipped with a terminal fringe; fruit 1/9 inch long, dark greenish to silvery, with several ivory-colored longitudinal stripes, the surface sparsely covered by minute short hairs, at the apex a tuft of stiff bristles somewhat shorter than the body of the fruit.

As in the case of all our other knapweeds, this species has been introduced. It is of European or West Asian origin, and probably reached western America through contaminated commercial seed. It is now common in various areas east of the Cascades.

DIFFUSE or SPREADING KNAPWEED

Centaurea diffusa Lam.

Annual or rarely biennial; stem ridged, rough, diffusely branched above, the whole plant at first thinly woolly; leaves grayish green, the lower several times divided, the upper once divided into narrow segments; heads slender, with pointed fringed bracts; flowers purple, rose or white; fruit 1/8 inch long, brownish or grayish with several pale longitudinal stripes, the apex with few or no bristles.

This species apparently originated in eastern Europe or adjacent Asia, and considerably resembles Spotted Knapweed. However, the spreading pointed bracts, without black tips readily separate it from that species. Diffuse Knapweed apparently entered most of our area even later than Spotted Knapweed, but is now becoming troublesome and is reaching out into new territory rather rapidly.

YELLOW STARTHISTLE
Centaurea solstitialis
a. upper branches x 2/3
b. lower leaf x 2/3
c. fruit x 8

MALTA STARTHISTLE
Centaurea melitensis
d. upper branches x 2/3
e. lower leaf x 2/3
f. fruit x 8

MALTA STARTHISTLE

Centaurea melitensis L.

Annual or biennial; stems generally several to many from base of the plant, 2 to 4 feet long, ridged, winged from prolonged leaf bases, the entire plant grayish green; lower leaves clustered, long, lobed, upper leaves narrow, often toothed; heads white-woolly in bud, often smooth later, the bracts tipped with sharp branching spines, the spines 1/6 to 1/3 inch long, sometimes purplish; flowers yellow, minutely glandular; fruit 1/8 inch long, olive gray, more or less striped longitudinally, bearing at the apex a tuft of grayish white bristles nearly equalling the body of the fruit in length.

A native of the Mediterranean region, this species has found its way into America and has become a pest in grainfields and pastures of California. Spreading northward, it is now occasionally encountered in western Oregon.

YELLOW STARTHISTLE

Centaurea solstitialis L.

Yellow Starthistle is, like Malta Starthistle, a yellow-flowered member of the Knapweed group. The two species somewhat resemble each other but Yellow Starthistle typically has larger heads, longer stiffer coarser spines, the flowers are not glandular, and the tuft of fruit bristles is silvery. This species probably is likewise a native of the Mediterranean region, and its distribution in western American has followed a route somewhat paralleling that of the preceding species. However, although not yet abundant except perhaps locally, it has spread over a wider range and is known in southwestern Idaho as well as Oregon.

RUSSIAN KNAPWEED

Centaurea repens L.

Perennial from deep-seated woody rootstocks; stems 1 1/2 to 3 feet tall, one to many from the base, ridged, leafy, the entire plant at first more or less white-woolly; lower leaves long, generally lobed, upper leaves smaller, toothed or entire; flowering branches many, each terminating in a head; heads narrow, usually less than 1/2 inch wide at the apex, the bracts whitish-papery; flowers pink to lavender; fruit whitish, obscurely striped, about 1/8 inch long, tipped at first by a tuft of long and short bristles, these falling at maturity.

This most dreaded of all knapweeds came originally from southern Russia and adjacent Asia, and probably was introduced into America in commercial seed. Within recent years it has become established chiefly east of the Cascades where it is troublesome in fields and along irrigation ditches, and where, in wastelands, it is developing sources of further infestation.

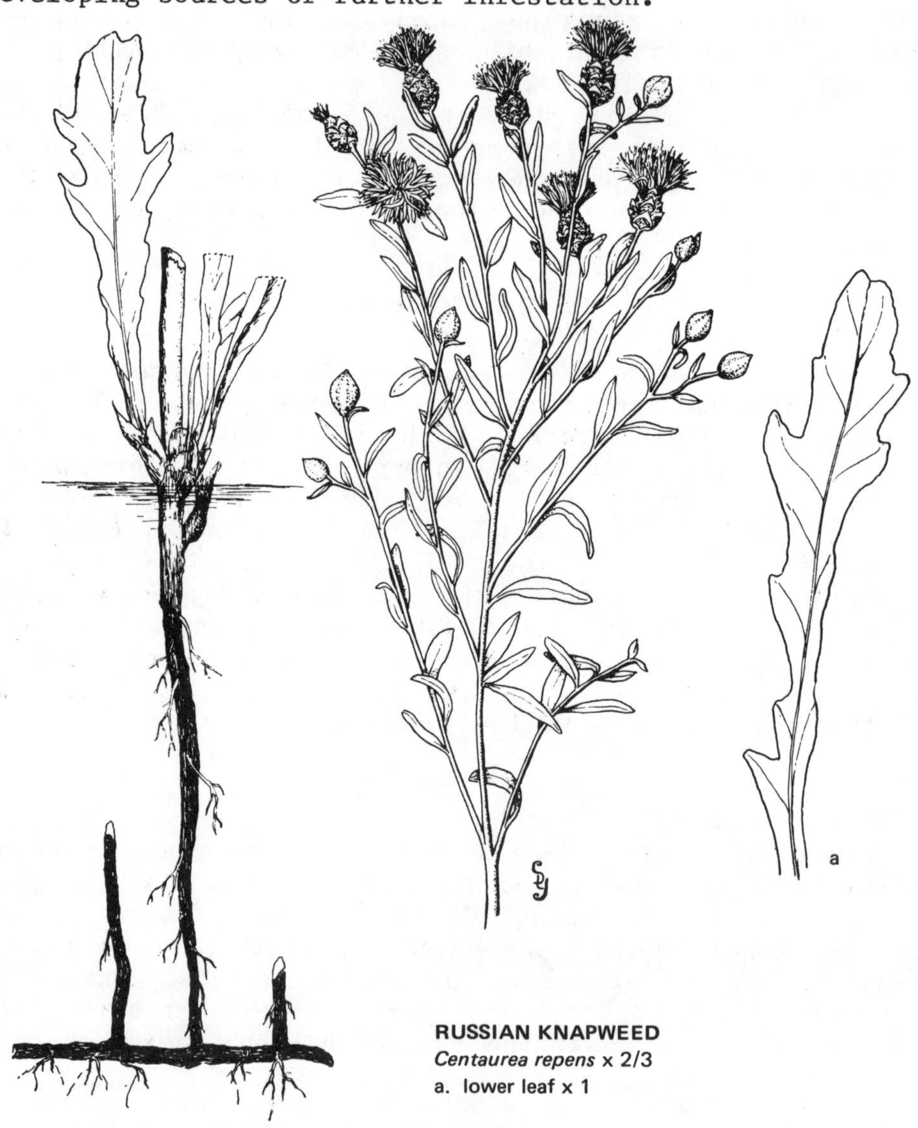

RUSSIAN KNAPWEED
Centaurea repens x 2/3
a. lower leaf x 1

CORNFLOWER or BACHELOR BUTTON

Centaurea cyanus L.

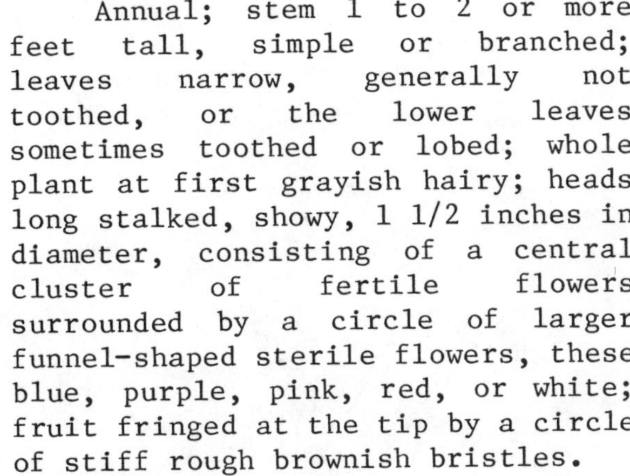

Annual; stem 1 to 2 or more feet tall, simple or branched; leaves narrow, generally not toothed, or the lower leaves sometimes toothed or lobed; whole plant at first grayish hairy; heads long stalked, showy, 1 1/2 inches in diameter, consisting of a central cluster of fertile flowers surrounded by a circle of larger funnel-shaped sterile flowers, these blue, purple, pink, red, or white; fruit fringed at the tip by a circle of stiff rough brownish bristles.

Although Cornflower is only an annual, its free seeding renders it difficult to control since its fruits with their circles of stiff bristles are equipped for wide dispersal. As a rule, the growing season of the plant is of short duration in the spring but spring cutting may result in autumn flowering and seeding. Scattered plants may be seen during almost any month of the year in western Oregon.

The plant is a native of the Mediterranean region but is said to have spread to nearly every part of the world, undoubtedly having been introduced in many areas as an ornamental garden species. It is better known west than east of the Cascades, and is commonly seen as a wayside and grainfield weed.

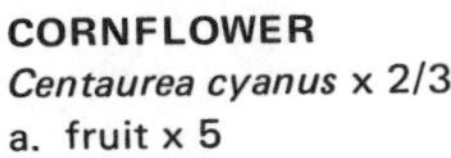

CORNFLOWER
Centaurea cyanus x 2/3
a. fruit x 5

BULL THISTLE
Cirsium vulgare
a. flowering stem x 2/3
b. leaf x 2/3
c. fruit x 6

PALOUSE THISTLE
Cirsium brevifolium
d. head x 2/3
e. leaf x 2/3
f. involucral bract x 4
g. fruit x 6

BULL THISTLE

Cirsium vulgare (Savi) Tenore

Biennial; stem 2 to 5 feet tall, erect, stout, branching, ridged and winged, more or less loosely woolly at first; basal leaves 1 foot or less long, deeply lobed, the lobes spine-tipped, upper side of the leaf covered by stiff hairs appressed to the surface and pointing toward the apex, under side loosely woolly, terminal lobe of the leaf elongated, lance-shaped; stem leaves smaller than the basal, similar in shape, clasping the stem and continuing below the attachment to form prickly wings; heads 1 1/2 to 2 or more inches high and broad, more or less clustered at the ends of branches; flowers reddish purple, the head enclosed by numerous narrow spine-tipped bracts; fruit somewhat flattened, 1/8 to 1/3 inch long, grayish or yellowish, with narrow broken dark lines; fruit at first tipped by a circle of plume-like white hairs an inch or more long forming the familiar "thistle down."

Although Bull Thistle has long been naturalized in the Pacific Northwest, it is of Old World origin, having come to us in impure seed or by other accidental means. In like manner it has spread to most other continents. It is common and abundant in pastures, fence-rows, farmyards, burns, clearings, and wasteland generally.

Several native thistles occur within our limits, but most of these are unimportant as weeds.

PALOUSE THISTLE

Cirsium brevifolium Nutt.
(=*C. palousense* Piper)

Biennial or sometimes perennial; stems 2 or more feet tall; leaves grayish woolly on the underside, less so on the upper side, the basal leaves long, comparatively narrow, cut into many lobes or divisions, these abundantly sharp-spiny; flowers white, the bracts of the head narrow with a central narrow dark gland, the outer bracts tipped by weak short spines bent outward at right angles.

This thistle occurs typically in southeastern Washington and adjacent Idaho and Oregon, appearing sporadically southward, its southernmost known limit being Lake County, Oregon. It is not a vicious weed but is causing concern in a few areas.

UTAH THISTLE
Cirsium utahense x 2/3
a. involucral bract x 4
b. fruit x 6

UTAH THISTLE

Cirsium utahense Petr.

Perennial, with a woody often branching underground system capable, under favorable conditions, of sending up new plants and establishing colonies; stem 1 to 2 or more feet tall, ridged, thinly white-woolly, branched or unbranched; leaves generally not longer than 6 or 7 inches, often shorter, narrow, lobed and toothed, margined with stiff sharp slender yellow prickles, these from 1/4 to nearly 1 inch in length, the spiny margins extending down the stem below the leaf bases, lower surface of leaves white-woolly, upper surface thinly so; heads borne singly or sometimes in 2's or 3's at ends of the branches and main stem, heads generally 1 1/2 to 2 inches in diameter at maturity, the bracts rather broad, often dark colored, ending in stiff spreading yellow spines; flowers white to light purple; fruit smooth, somewhat ridged.

This thistle, which for many years has been known in southwestern America, has come to our attention in Oregon in comparatively recent years. In Eastern Oregon it has spread to some extent and, because of the colonizing tendency of its underground system, has the potentialities of a pest. It occurs also in southeastern Washington and Idaho.

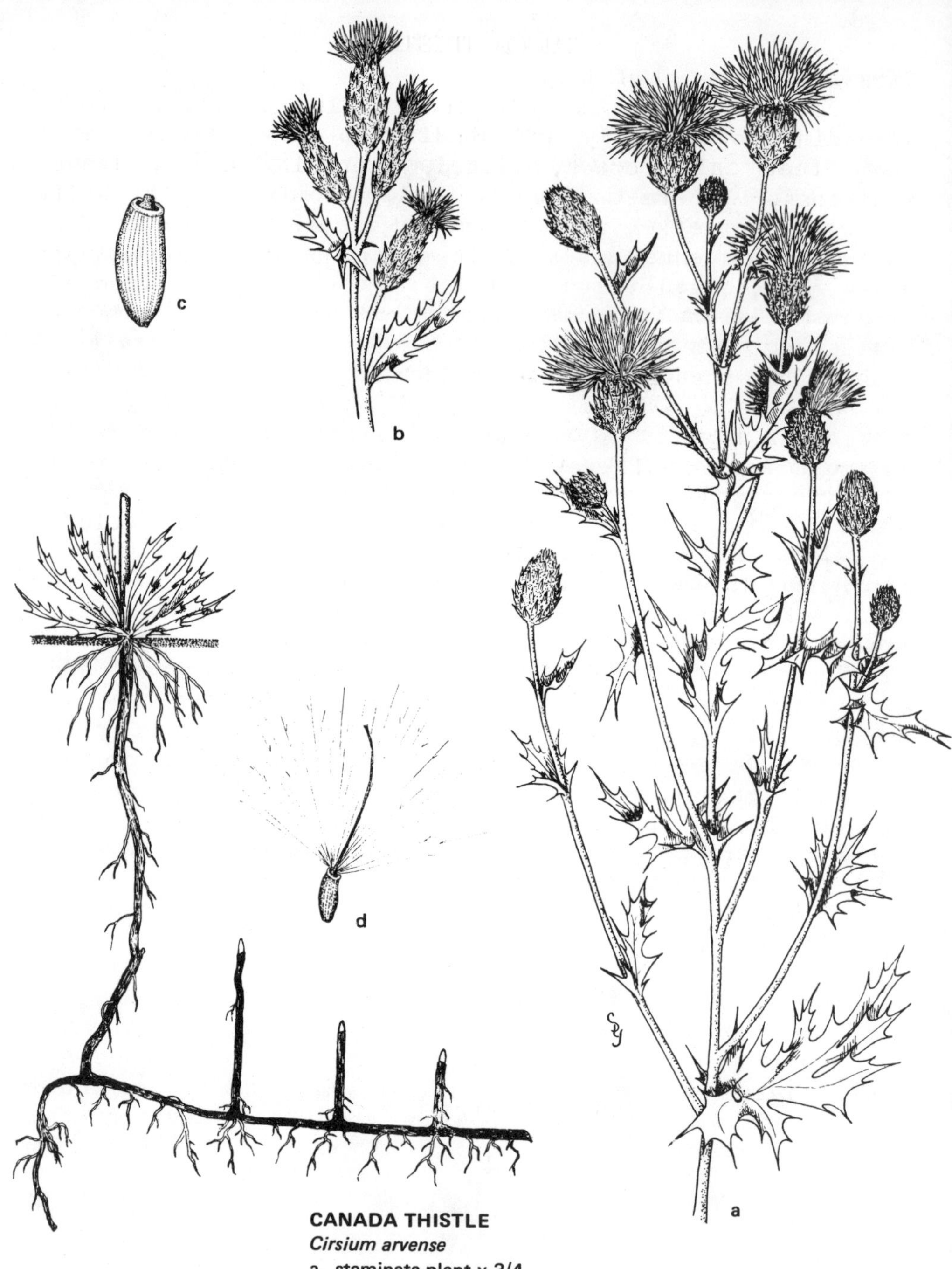

CANADA THISTLE
Cirsium arvense
a. staminate plant x 3/4
b. pistillate plant x 3/4
c. fruit after pappus has fallen x 6
d. fruit x 2

CANADA THISTLE

Cirsium arvense (L.) Scop.

Perennial by a deep-seated complex system of roots spreading horizontally and sending up new stems; stems 1 to 4 feet tall, erect, ridged, branching above; leaves stalkless, generally deeply lobed, edged with stiff yellowish spines, 2 rows of these generally extending down the stem for some distance below the base of each leaf; heads many, purple, small, borne at tips of the branches, often in large clusters; fruit 1/8 inch long, somewhat flattened, brownish, with at first an apical circle of long hairs, these eventually falling.

Although commonly called "Canada" Thistle, this species in reality is a native of southeastern Europe and adjacent Asia and, even before the discovery of America, was recognized on the continent of Europe as a pernicious weed. It was introduced into Canada by early settlers, probably in contaminated seed, and appeared shortly afterward in New England, though perhaps directly from Europe rather than from Canada.

As early as 1795, legislation was enacted against it in America. However, it continued to spread and, reinforced by numerous subsequent introductions, now covers a large part of both Canada and the northern United States. No continent is now free of it and in all infested agricultural sections it is rated one of the most serious threats to crop production.

By means of its extensive branching root system, which functions both in food storage and expansion, it is able to cover large areas from a single point of infestation. Burrowing through the soil horizontally, as the root tips elongate these laterals send up new stems at intervals. Breaking or cutting these roots by cultivation does not kill them. Instead, each severed portion is able to establish new plants, and the plow may thus even aid their distribution. The depth to which the underground system penetrates depends upon the moisture content of the soil, its richness, texture, and tilth. Tillage tends to send the roots downward and the layer of horizontally spreading and interlacing roots may lie at a depth of 1 1/2 to 2 1/2 feet. In long uncultivated land it is usually much nearer the surface.

The flower heads are of two kinds, staminate and pistillate. Typically each plant bears only one kind of flower head, and a colony produced by underground propagation of a single plant will therefore be entirely staminate or pistillate. Since both kinds are necessary for seed production, individuals and whole colonies are often sterile.

Canada Thistle is widely variable in certain characters. A form occasionally seen has nonlobed and spineless leaves. However, this species is easily recognized by the small, clustered heads, pistillate and staminate flowers on different plants, and the perennial, underground system of propagation.

In milder parts of the Northwest, Canada Thistle often overwinters with a rosette of green leaves ready to begin work with the first favorable days of spring. The manufacture of food is thus started at the earliest possible moment, to be used in growth, seed production, and underground expansion necessary to colonization. Enormous quantities of food are stored below ground for future use, and every green shoot above ground not only adds to this store but hinders control.

MILK THISTLE

Silybum marianum Gaertn.

Biennial or winter annual; stem 2 to 5 feet tall, stout, ridged, generally branching, erect, slightly cobwebby; leaves sometimes reaching 1 1/2 feet in length, broad, lobed, clasping the stem with ear-like lobes at the base, spiny-margined, with white marbling along the veins; head thistle-like, with leathery spine-tipped bracts; flowers red-purple; fruit 1/4 inch long, shining, sometimes brown mottled, tipped by a tuft of shining white minutely barbed hairs, this tuft falling as a unit at maturity.

This spectacular native of Europe is found rather abundantly in southwestern Oregon, and occurs sporadically in other parts of the area. It spreads thistle-wise by plume-equipped fruits.

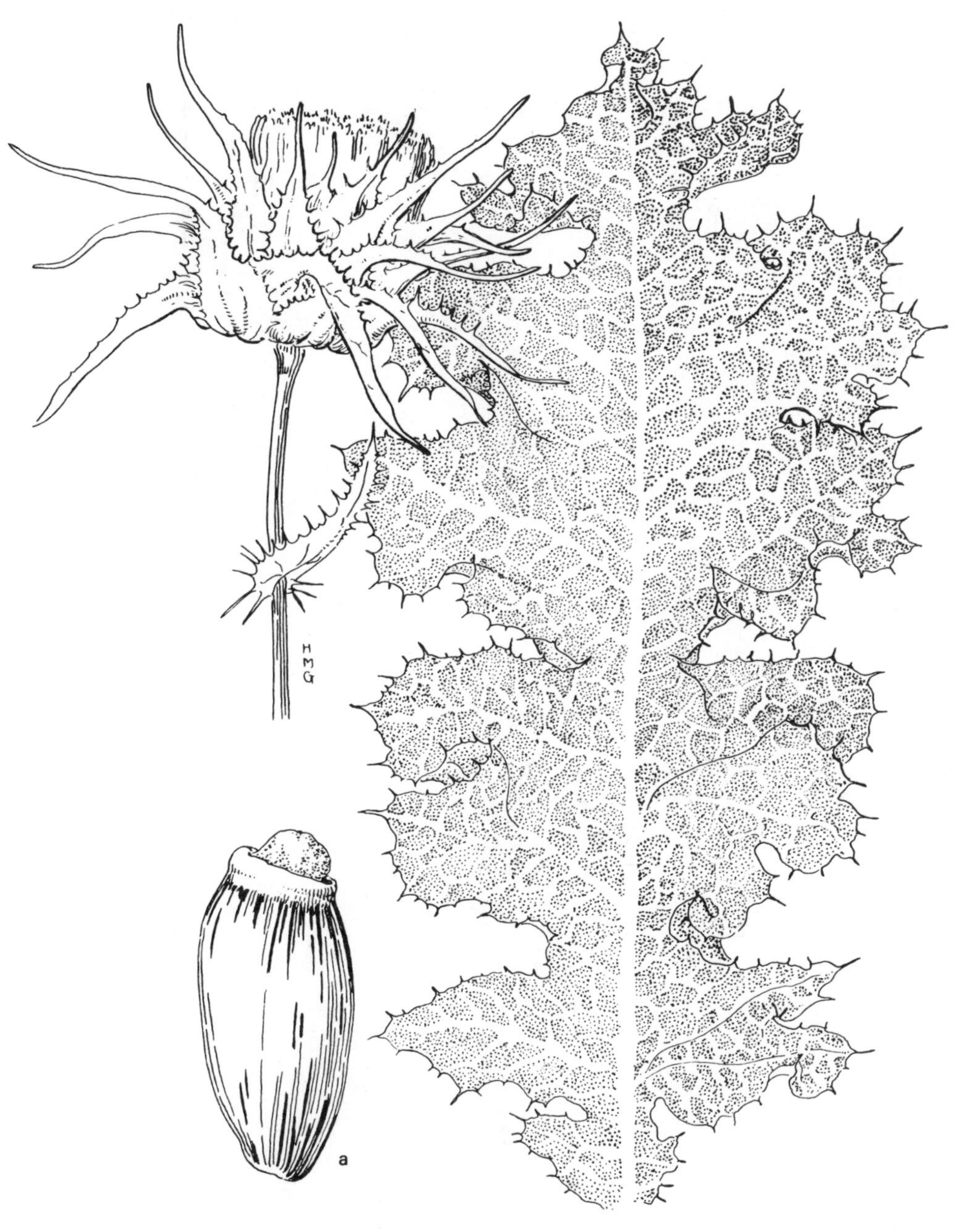

MILK THISTLE
Silybum marianum x 2/3
a. fruit after pappus has fallen x 6

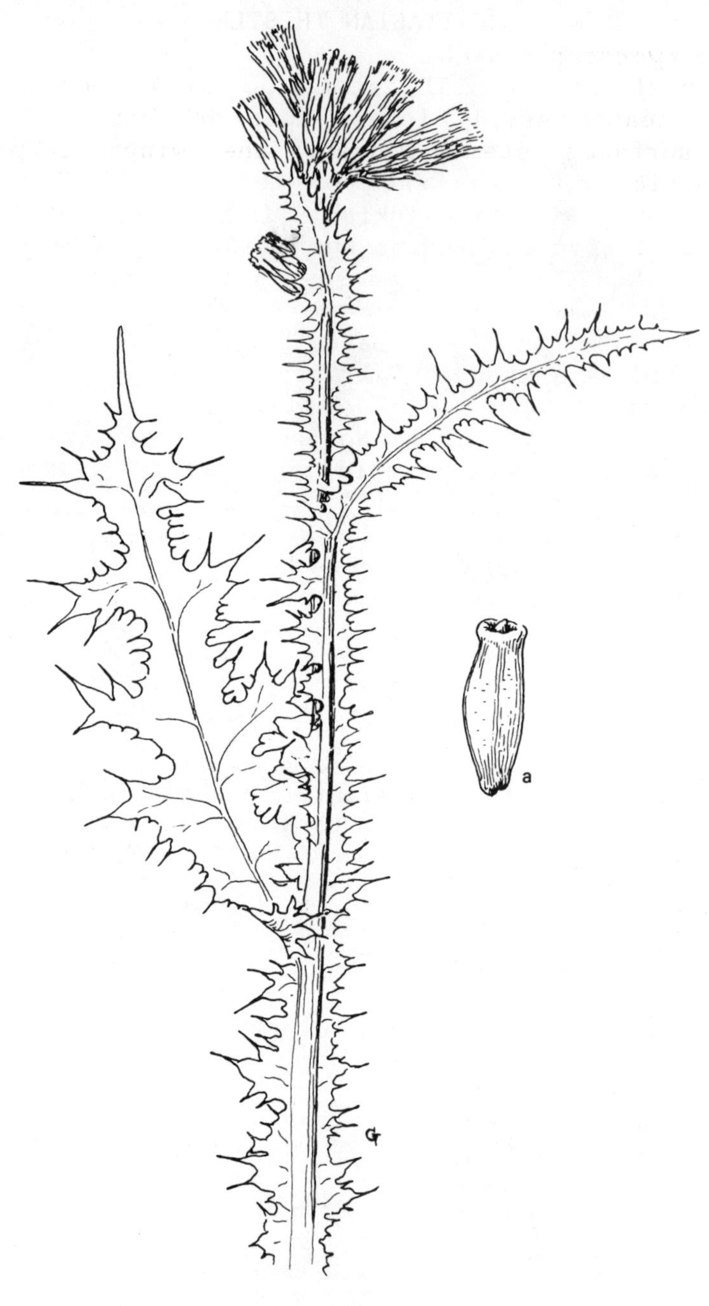

ITALIAN THISTLE
Carduus pycnocephalus x 4/5
a. fruit x 5

ITALIAN THISTLE

Carduus pycnocephalus L.

Annual or biennial; stems up to 4 feet or more in height; leaves deeply lobed, more or less woolly on the under surface; stems winged, the wings formed by a continuation down the stem of the prickly leaf margins; flowers purplish or pinkish, in small, slender heads mostly 2- to 5-clustered at the ends of the branches.

This Mediterranean species is found only occasionally within our boundaries, from the southern Willamette Valley southward along the coast, it is, however, well established as a weed in California.

Two other species of *Carduus* are also occasionally reported from the Pacific Northwest: *Carduus tenuiflorus* Curtis (Slenderflower Thistle) and *Carduus nutans* L. (Musk Thistle). *Carduus tenuiflorus* is similar to *Carduus pycnocephalus* but usually with more than 5 heads to a cluster; it too occurs along the southern Oregon coast. *Carduus nutans*, has larger, broader, and solitary heads; it is known from Washington and northern Oregon, both east and west of the Cascades.

SCOTCH THISTLE

Onopordum acanthium L.

Similar to the genus *Cirsium*; spiny biennial up to 8 feet high; stems with broad spiny wings; leaves white-woolly, toothed or lobed, spiny, up to 2 feet long and 1 foot wide, but usually much smaller; heads numerous, 1 to 2 inches in diameter, the bracts all spine-tipped; flowers purple; fruits about 3/16 inch long, slightly flattened and wrinkled, tipped by slender bristles.

Native of Europe, sporadically established throughout the United States. In our region, most commonly reported from roadsides and wasteland along the Snake River and in southeastern Oregon, but occasionally from west of the Cascades.

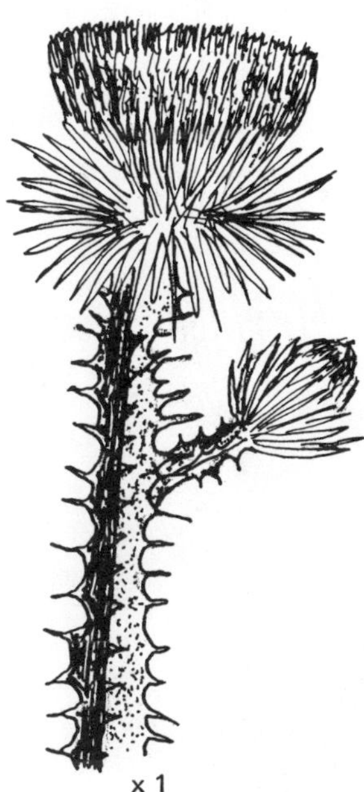

BLESSED THISTLE

Cnicus benedictus L.

Annual; stem widely branching, 1 to 2 feet tall, stout, ridged, often reddish, cobwebby; lower leaves lobed, with prickly margins, stalked, the stalks winged, upper leaves stalkless, broadest at the base, clasping the stem; uppermost leaves clustered beneath the head and more or less enclosing it; all leaves roughened, veiny and prickly-margined; flowers yellow, in a compact head surrounded by yellowish bracts, each bract ending in a long brownish horny branched spine, all enclosed in the cluster of upper green leaves.

Although not so abundant as to have become a menace, this plant is frequently encountered and reported. It is a native of the Mediterranean and the Caucasus regions.

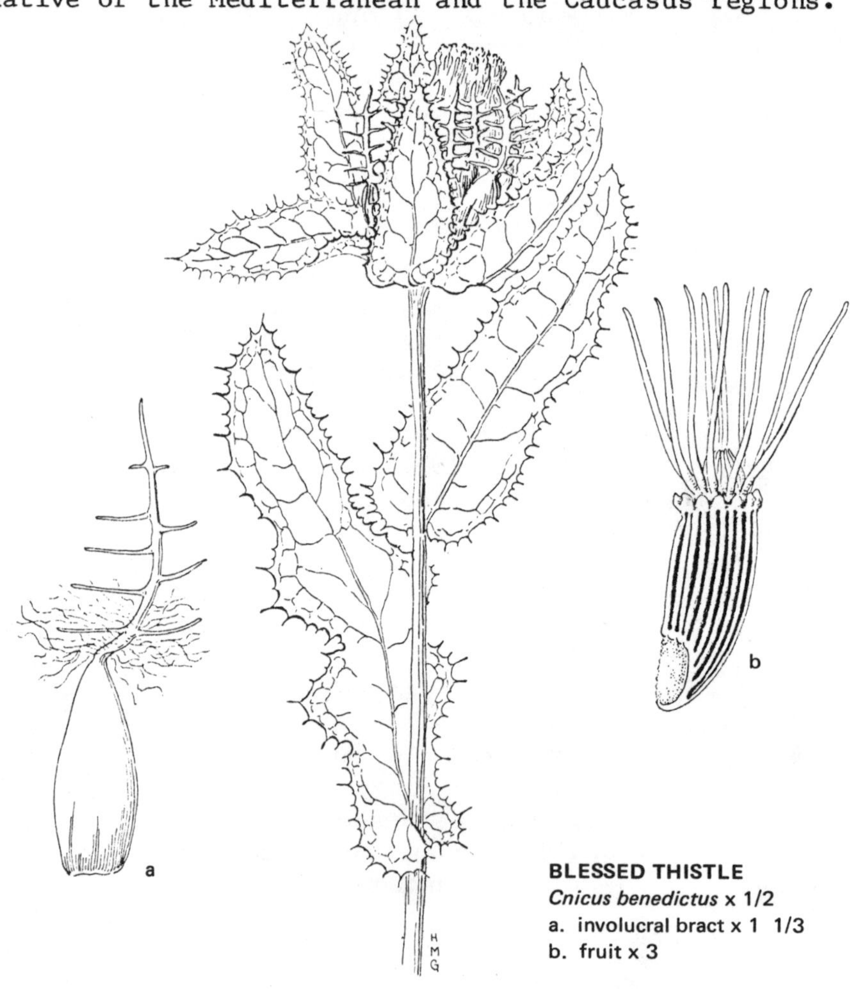

BLESSED THISTLE
Cnicus benedictus x 1/2
a. involucral bract x 1 1/3
b. fruit x 3

COMMON YARROW

Achillea millefolium L.

x 9/10

Perennial; stem stout, erect, generally unbranched below; stem and leaves nearly smooth to hairy or woolly; leaves several times divided into numerous small segments, the entire leaf appearing fern-like; plant possessing a distinct aromatic odor when bruised; flowers white or rarely pink; flower heads small, arranged in a large, nearly flat-topped cluster.

This is a widely distributed and highly variable species with several varieties represented in our region; var. *lanulosa* (Nutt.) Piper being the most common.

Common Yarrow is a well-known weed of roadsides, waste places, pastures and lawns. Under good pasturage conditions, this species is generally not readily eaten by stock.

MAYWEED
Anthemis cotula x 4/5
a-b. fruits x 12

MAYWEED or DOG FENNEL

Anthemis cotula L.

Annual; plant generally bushy-branched, ill-smelling, slightly hairy, 1/2 to 2 feet tall; leaves several times divided into narrow segments; heads generally many, 1 to 1 1/2 inches in diameter, long-stalked, borne at ends of the branches and in leaf axils, rays 12 (or slightly more or fewer), white; center of the head rounded at first becoming coneshaped; fruit ribbed, warty.

A native of Europe, Mayweed has followed man over the world and now is known on every continent. It is a strong-smelling weed of waste land and cultivated fields, and can adapt itself to widely differing conditions of soil, drainage, exposure, and climate. It is common and abundant in many barn lots, chicken runs, and in neglected borders and corners about buildings.

In addition to being valueless in itself and crowding out other plants, Mayweed in some cases produces contact irritation somewhat similar to that induced by Poison Oak, though generally of shorter duration; and, in sensitive persons, the odor produces irritation of the nostrils, accompanied by sneezing. The mouths and noses of animals eating hay contaminated by Mayweed may be burned and blistered. When these plants are eaten by cows a strong disagreeable flavor is evident in their dairy products.

A related introduced species commonly called Field Chamomile (*Anthemis arvensis* L.) occasionally occurs in our region. Similar in general appearance to Mayweed, it differs in usually larger heads, ribbed but not warty fruit and the lack of a disagreeable odor.

PINEAPPLE WEED or RAYLESS DOG FENNEL

Matricaria matricarioides (Less.) Porter

Annual; strongly-scented, erect, 2 inches to 1 1/2 feet tall; leaves greatly divided into very short narrow segments; heads rounded-conical, without rays; fruit ridged, bearing a microscopic tooth-like point at the apex.

Although this species is now common in Europe, it is believed to have originated in western America whence it spread to other continents, thus reversing the usual direction of travel followed by our weeds. It shares with

Mayweed, which came to us from Europe, the distinction of being one of our most widespread, abundant and tenacious weeds of roadsides, vacant lots, gardens, clearings, burns, and other places where it may gain a foothold. Dry, hard-packed paths and trampled farm yards often exhibit solid stands of one or both of these two species, for competition here is low. Such areas furnish centers of distribution to gardens and cultivated fields.

PINEAPPLE WEED
Matricaria matricarioides x 4/5
a. fruit x 12

TANSY or GARDEN TANSY

Tanacetum vulgare L.

Aromatic perennial; stems 1 1/2 to 6 feet tall; leaves divided into numerous narrow, toothed segments; heads numerous, about 1/4 to 3/8 inch in diameter, cup-shaped, golden yellow, the bracts narrow and overlapping; fruit 5-ridged.

Native of Europe and commonly cultivated, escaped and well established along roadsides and stream banks; most common in the Willamette Valley, but occasional reports come from other areas, including localities east of Cascades.

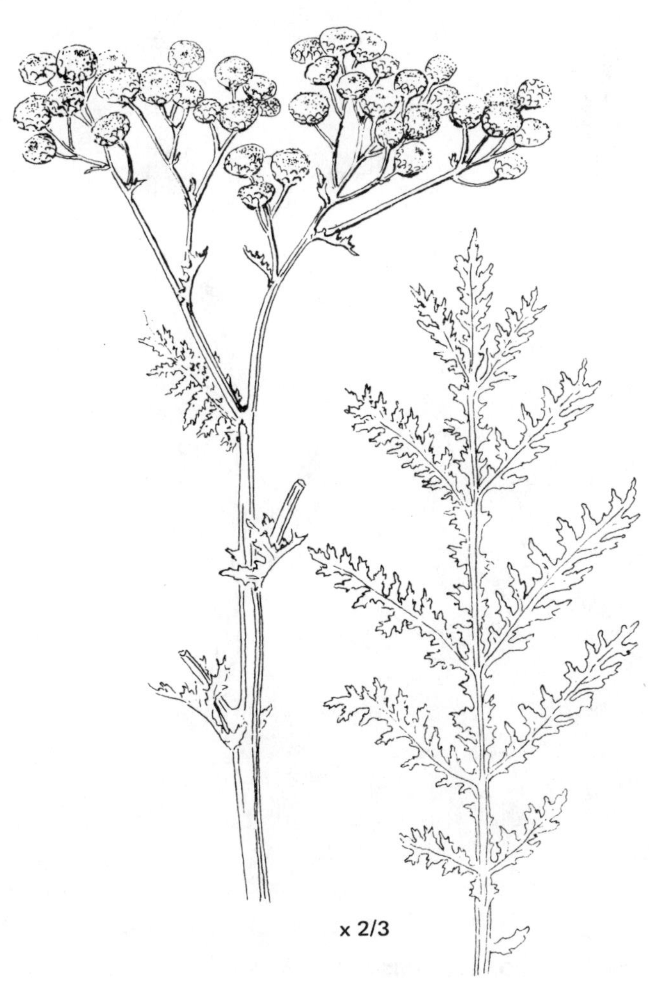

OXEYE DAISY
Chrysanthemum leucanthemum x 2/3
a. fruit x 12

OXEYE DAISY or MARGUERITE

Chrysanthemum leucanthemum L.

Perennial, spreading by short rootstocks; stems minutely ridged, sometimes purplish, one or several from the base of the plant, erect, 1 to 3 feet tall, simple or branching above; leaves of ours coarsely toothed or deeply lobed, the lower leaves long-stalked, the uppermost leaves more or less clasping the stem by a broad fringed or toothed base; flower heads at the ends of long stems bearing a few reduced leaves; heads 1 1/2 to 2 inches in diameter, with many white rays and a yellow button-like central portion bearing minute crowded disk flowers; fruit 1/16 inch long, dark, with 10 whitish ribs.

Oxeye Daisy, a native of Europe, was first introduced into America as a garden plant, but early escaped cultivation and has long been known as a wayside weed. It is particularly common west of the Cascades in Oregon, and is abundant and attractive on roadsides, pastures and meadows of the Willamette Valley. Its spread in pastures is stimulated by the fact that for the most part it is ignored by grazing animals which feed, instead, upon more desirable plants such as grasses. The undisturbed Oxeye Daisy is thus free to seed and scatter, while the closely-cropped grasses are held in check. The minute fruits may form an impurity in commercial crop seed.

The common western form of Oxeye Daisy with more coarsely toothed or lobed lower leaves than are typical for the species, has commonly been known as variety *pinnatifidum* Lecoq. & Lemotte (Field Oxeye Daisy).

BIG SAGEBRUSH

Artemisia tridentata Nutt.

Perennial; stems branching and shrubby, 1 to 10 or more feet in height; entire plant grayish-hairy; leaves small, broadest near the apex, entire or more often 3- to 5-toothed or divided at the apex; heads small, in large clusters, whitish or grayish-hairy or nearly smooth.

This is the common sagebrush of the plains east of the Cascades. Several varieties or subspecies are recognized by most botanists. To the extent that it has spread over range lands once bountifully supplied with nutritious perennial grasses, it constitutes a weed. Overgrazing has stimulated its rapid increase and, though it too is grazed, its returns are probably small in comparison with the plants which it has replaced. All species of sagebrush, with their penetrating odor, are known to produce marked hay-fever reactions in many persons.

Several other species of sagebrush occur, principally east of the Cascades, though a few forms are not uncommon west of the mountains.

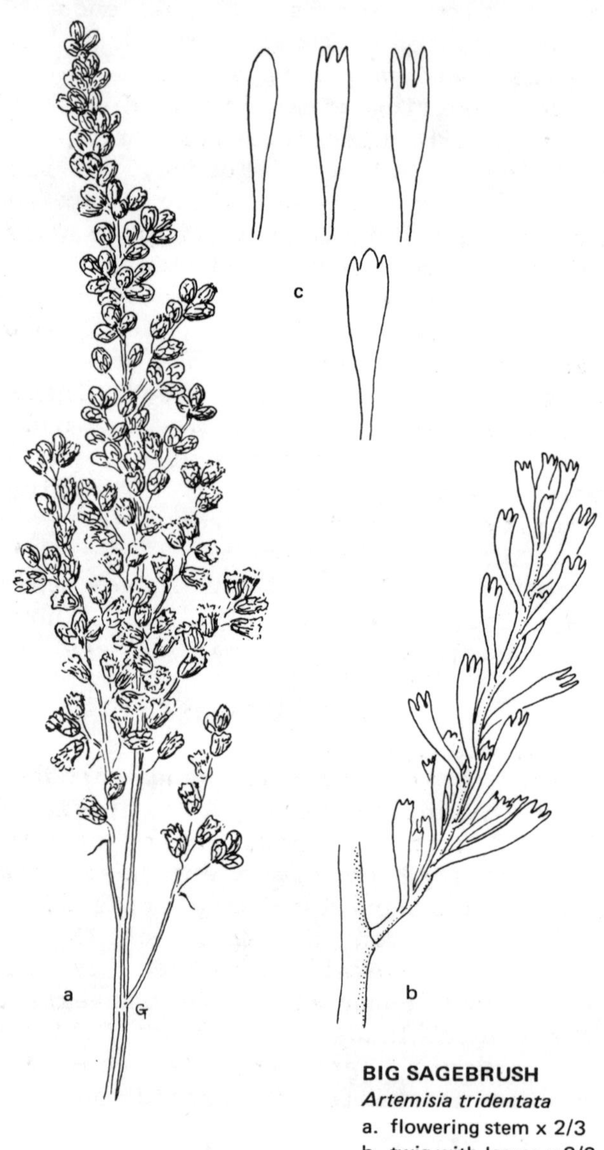

BIG SAGEBRUSH
Artemisia tridentata
a. flowering stem x 2/3
b. twig with leaves x 2/3
c. leaf variation x 1 1/4

PEARLY EVERLASTING

Anaphalis margaritacea (L.) B. & H.

Perennial, with a slender rootstock; stem 1 to 3 feet tall, erect, branched or unbranched, more or less white-woolly, leafy; leaves long, narrow, stalkless, densely white-woolly; heads small, in large terminal clusters with conspicuous white papery bracts.

This well-known native species occurs on both sides of the Cascades in one or another of its various forms. It is common in open woods or, associated with Fireweed and native Blackberry, is often abundant on old burns and clearings. It sometimes becomes a pasture weed, but rarely causes trouble in cultivated fields.

TANSY RAGWORT
Senecio jacobaea x 3/4
a. fruit x 8

TANSY RAGWORT

Senecio jacobaea L.

Biennial or perennial; strongly scented; stem 1 to 6 feet tall, erect, often widely branched above, generally rather stout, finely ridged, green or dark red, often thinly covered at first by cobwebby hairs; leaves divided into lobed and toothed segments, the terminal lobe generally much larger than the laterals; flowering heads typically numerous, each with 10 to 15 golden-yellow rays; fruits somewhat elongated, those of the marginal flowers smooth, of the central flowers short-hairy, all at first tipped by a tuft of long white hairs, these soon falling.

Although Tansy Ragwort has entered Oregon in comparatively recent years, it has spread so rapidly and proved so unwelcome that it now is probably one of our best known weeds. During its first year in an area, it sends up a thick cluster of dark green much-divided leaves. Early in the second year, if conditions are favorable, a stout stem arises, branches freely above, and typically produces a widely spreading somewhat flat-topped arrangement of numerous golden-yellow flower heads which are followed by masses of fruits equipped with tufts of white hairs somewhat resembling those of thistledown. These fruits are carried far and wide by the wind, until eventually they settle down to repeat the process in new areas.

Under some conditions, the plant behaves as a biennial dying at the end of the second season after producing an abundant crop of seed. Typically it is a perennial, continuing to produce seed each season.

The ability of the plant to accomplish regeneration from the roots, even broken remains of roots after the tops have been cut or pulled, adds to the difficulty of control. In the spring, when growth conditions are favorable, numerous young shoots may be seen emerging from such roots and developing into normal plants. Even plants whose tops have been killed by spraying sometimes regenerate from undamaged crowns, thus weakening control by this means.

Tansy Ragwort has the reputation of being poisonous to stock, especially cattle and horses.

COMMON GROUNDSEL

Senecio vulgaris L.

Annual; more or less hairy or loose-woolly or smooth; stem 1/2 to 1 1/3 feet tall, hollow, rather succulent, ridged, branched; leaves lobed and toothed; heads cylindrical, several to many, more or less clustered, with black-tipped bracts, rays none; fruit slender, ridged, somewhat short-hairy, tipped at first by a tuft of long silky white hairs.

Naturalized from Europe, this weed is now locally abundant in fields, waste places, and gardens.

WOODLAND GROUNDSEL

Senecio sylvaticus L.

Annual; plant with a nauseating odor when bruised; stem 1 to 3 1/2 feet tall, generally simple below, branched above, ridged, more or less loose-woolly, leafy; leaves deeply lobed or divided, largest near the base of the plant, becoming increasingly smaller upward; flowering branches generally many, spreading, or nearly erect; heads narrowly cylindrical, small, with narrow black-tipped bracts; fruit nearly 1/8 inch long, narrow, ridged longitudinally, covered by microscopic stiff hairs lying parallel to the surface, the fruit tipped at first by a cluster of long fine silky white hairs.

Woodland Groundsel, a native of Europe, apparently first appeared in Oregon on the southwestern coast, whence it has spread northward and inland up the river valleys until it is now abundant in much of the waste land west of the Cascades. It seeds as freely as thistles and also is dispersed by numerous parachute-like fruits which are carried far and wide by prevailing winds. The plant is of no known value in itself, is objectionable to stock, and is unsightly along the highways where it often succeeds in crowding out all desirable vegetation. It bears a slight resemblance to Tansy Ragwort and sometimes is mistaken for it. This fact is not too serious perhaps, since it too is a noxious weed.

In reality, only in its young stages can Woodland Groundsel be confused with Tansy Ragwort, and even then the differences are rather well marked. Since Woodland Groundsel is an annual, it does not form a large first-year cluster of basal leaves; the foliage is grayish-green; the leaves do not exhibit the large terminal lobe characteristic of most plants of Tansy Ragwort; and the flower heads are small, narrow, and inconspicucus, without rays. Plants of both species possess a strong disagreeable odor when bruised, but that of Woodland Groundsel is particularly nauseating and revolting, and in the fresh stage will readily distinguish the species from any other *Senecio* in our area.

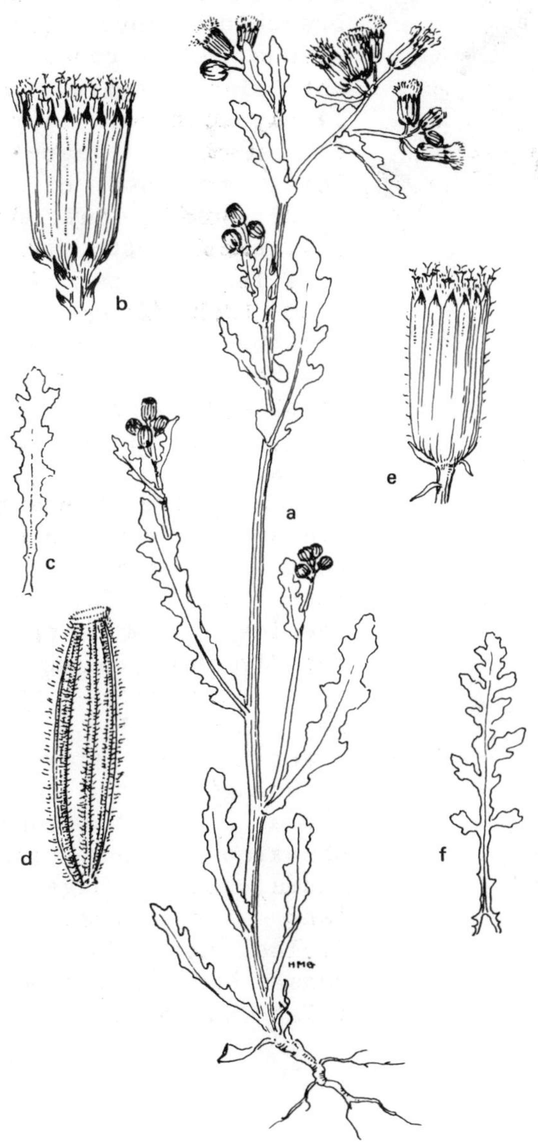

COMMON GROUNDSEL
Senecio vulgaris
a. plant x 2/3
b. head x 4
c. lower leaf x 2/3
d. fruit x 4

WOODLAND GROUNDSEL
Senecio sylvaticus
e. head x 4
f. lower leaf x 2/3

AUSTRALIAN BURNWEED or FIREWEED

Erechtites minima (Poir.) DC.
(*Erechtites prenanthoides* misapplied)

x 3/4

Annual; stem stiff, erect, 1 to 5 feet tall, ridged, often purplish or reddish, especially below, generally widely branching above; leaves slender, narrowed at both ends, sharply and finely toothed, clasping the stem with ear-like lobes at base of the blade; heads many, narrow, small, with slender parallel bracts; fruit 1/16 inch long, brown, 5-ribbed, slightly roughened, bearing at first a terminal tuft of long, very fine white hairs.

A native of Australia and believed to have been introduced first into California, this species has spread rapidly north along the coast of Oregon. It is an ill-smelling unattractive weed which is crowding out the native flora and becoming obnoxious along roadsides, burns, logged-off land, pastures and fields, not only on the coast but also now in inland valleys.

New Zealand Fireweed, *Erechtites arguta* DC., is similar in habit and in ability to reproduce itself rapidly, though at present it is less common in Oregon than the preceding species. It resembles Australian Burnweed in general but is often shorter, with less widely-spreading branches. The leaves are irregularly lobed, and the heads are white-woolly at the base. The fruit is slightly shorter and wider.

GLOSSARY

ACHENE: generalized term for a small dry one-seeded fruit not splitting open at maturity.

ALTERNATE: borne singly at a node, as leaves, or stamens not in front of petal lobes.

ANNUAL: a plant completing its life cycle in a single season.

ANTHER: the pollen bearing portion of the stamen.

APPRESSED: pressed flat against the surface.

AQUATIC: living in water or marshy places.

ASYMMETRICAL: with the two sides or halves unequal in size or shape.

AURICLE: a small ear-shaped appendage or lobe.

AWN: a slender, usually terminal, bristle.

AXIL: upper angle between leaf and stem.

AXILLARY: produced in the leaf axil.

AXIS: (of a stem, inflorescence, etc.) the central elongated column to which an organ or its parts are attached.

BEAK: the narrowed tip of a fruit or seed.

BERRY: a type of fruit which is fleshy throughout.

BIENNIAL: a plant which requires two years to complete its life cycle, then dies.

BLADE: the more or less expanded portion of a leaf or a petal.

BRACT: more or less reduced or modified leaf associated with a flower or flower cluster.

BULB: a short underground shoot with modified thickened leaves, eg. onion.

BULBIL: a small bulb-like structure that may be aerial or underground; sometimes produced in the inflorescence in place of flowers.

CALYX: all the sepals of the flower collectively; the outer whorl of flower parts.

CAPITATE: head-like.

CAPSULE: a dry fruit derived from a compound pistil and splitting open at maturity to release the seeds.

CLAW: narrow stalk-like base of a sepal or a petal.

COMPOUND LEAF: one with two or more leaflets.

COROLLA: all the petals of a flower collectively; the second whorl of floral parts (starting from the outside).

DISK FLOWER: tubular-shaped flower of the Compositae, borne in the center of the head when strap-shaped flowers are also present.

DORSAL: pertaining to the back or outer surface of a leaf, petal, etc.

ELLIPTICAL: widest near the center and curving toward both ends.

ENTIRE: not toothed or interrupted.

FERTILE: bearing reproductive structures.

FILAMENT: the stalk supporting the anther of a stamen.

FLESHY: thick and usually juicy.

FLORET: small individual flower of an inflorescence, particularly of Gramineae and Compositae.

FRUIT: the ripened seed-bearing part of the pistil together with any parts which develop with it as a unit.

FROND: term applied to the leaves of ferns.

GENUS: the main division of the plant kingdom below family and above species.

GLAND: a structure on the surface of an organ or at the end of a hair which produces a sticky substance.

GLOBOSE: round, spherical.

GLUME: one of the two empty scales or bracts at the base of a spikelet in Gramineae.

HEAD: type of inflorescence in which the flowers are not stalked and are closely crowded together on a shortened axis; the outer flowers maturing first.

HERB: a non-woody plant, dying back to the ground at the end of the growing season.

IMPERFECT FLOWER: one with either stamens or pistil, but not both; unisexual.

INFERIOR OVARY: where the bases of other floral parts are fused with the ovary.

INFLORESCENCE: the arrangement of flowers, or the flower cluster.

INVOLUCRE: a number of bracts beneath an inflorescence, as in Compositae.

IRREGULAR FLOWER: one in which the petals (or sepals) are different in size or shape.

LANCEOLATE: narrow, broadest near the base, tapering to the apex.

LEAFLET: leaf-like unit of a compound leaf.

LEMMA: one of the pair of bracts subtending the individual flower in the Gramineae.

LIGULE: the appendage on the inner side of the leaf at the junction of the sheath and the blade, as in Gramineae; also applied to the flattened portion of the ray flower in the Compositae.

LINEAR: narrow and with parallel sides.

MIDRIB: central vein of a leaf blade, bract, sepal, petal, etc.

NETTED: the ultimate veins forming a network.

NODE: joint; the point on the stem from which leaves and branches arise.

NUTLET: a one-seeded division of a specialized fruit from a lobed ovary, dry, not splitting at maturity; occurs in two's or multiples of two's in each flower.

OPPOSITE: borne in pairs, directly across from each other at a node, as in leaves, or borne directly in front of another organ, as in stamens opposite the petals.

OVARY: enlarged basal portion of the pistil enclosing the ovules or developing seeds.

OVATE: egg-shaped in outline.

PALMATE: leaflets, lobes or veins radiating from one point.

PANICLE: a branched inflorescence in which the lower flowers open first.

PAPPUS: the modified calyx of the Compositae.

PARALLEL-VEINED: veins running side-by-side, not meeting, except rarely at the apex or near the margin.

PEBBLED: minutely roughened.

PERENNIAL: a plant which continues to live year after year.

PERFECT FLOWER: one with both stamens and pistil(s).

PERIANTH: the two outer whorls of floral parts, the sepals and petals collectively.

PETAL: a member of the second set of floral parts (corolla).

PINNATE: lobes, leaflets or veins arising along the sides of a common axis.

PISTIL: the seed-producing portion of the flower, usually differentiated into ovary, style and stigma.

PISTILLATE FLOWER: a flower bearing a pistil or several pistils, but without stamens.

PLUMOSE: feathery.

PROSTRATE: flat on the ground or substrate.

RACEME: a more or less elongated inflorescence with stalked flowers and with the lower flowers maturing first.

RAY: one of the strap-shaped flowers in the outer whorl of flowers in the Compositae; also applied to one of the branches in an umbel.

RECEPTACLE: the end of the stalk to which the other floral parts are attached; in the Compositae, the structure where the flowers of the head are attached.

REGULAR FLOWER: one in which the petals (or sepals) are the same size and shape.

ROOTSTOCK: an underground root-like structure giving rise to new plants, usually an underground stem (rhizome).

ROSETTE: a cluster or circle of organs (often leaves) usually in a basal position.

SEPAL: a single member of the outermost whorl of floral parts (calyx); typically green.

SESSILE: not stalked.

SHEATH: lower part of the leaf enclosing the stem, as in Gramineae, Cyperaceae and Juncaceae.

SHEATHING: wrapping around; partly or wholly surrounding.

SPECIES: main division of a genus.

SPIKE: a more or less elongated inflorescence, similar to a raceme, except with sessile (non-stalked) flowers.

SPIKELET: in Gramineae, a floret or florets plus the glumes; also used to describe a segment of the inflorescence in other families, eg. Cyperaceae.

STAMEN: pollen producing structure of the flower plus its stalk; differentiated into anther and filament.

STAMINATE FLOWER: flower with stamens but without a pistil.

STERILE: without sexual reproductive structures; infertile.

STIGMA: part of the pistil fitted for receiving pollen, usually the tip.

STIPULE: one of a pair of basal appendages found on many leaves.

STOLON: an elongated creeping stem on the surface of the ground, rooting at the nodes and sending up new plants.

STYLE: slender stalk connecting the stigma to the ovary.

SUCCULENT: fleshy and juicy.

SUPERIOR OVARY: where the floral parts are not fused with the ovary.

THREE-RANKED LEAVES: borne in three vertical rows on the stem, as in the Cyperaceae.

TOOTH: a pointed structure arising along the edge of an organ, typically a leaf.

TUBER: a short fleshy modified stem generally borne at the end of a rhizome.

TUBERCLE: a small swelling or projection.

TWO-RANKED LEAVES: borne in two vertical rows on the stem, as in Gramineae.

UMBEL: a flower cluster in which the individual flower stalks arise from a single point.

UNISEXUAL FLOWER: one with either stamens or pistil(s), but not both; imperfect.

VEGETATIVE: pertaining to leaves, stems and roots, as contrasted to flowers.

VEINS: the vascular bundles, as in a leaf.

WHORLED: borne in a circle, more than two from a single node, as in leaves.

WING: a thin, flat projection.

INDEX